SURFACE TRANSPORTATION CONGESTION

SURFACE TRANSPORTATION CONGESTION

WILLIAM J. MALLETT

Novinka Books
New York

Copyright © 2008 by Nova Science Publishers, Inc.

All rights reserved. No part of this book may be reproduced, stored in a retrieval system or transmitted in any form or by any means: electronic, electrostatic, magnetic, tape, mechanical photocopying, recording or otherwise without the written permission of the Publisher.

For permission to use material from this book please contact us:
Telephone 631-231-7269; Fax 631-231-8175
Web Site: http://www.novapublishers.com

NOTICE TO THE READER

The Publisher has taken reasonable care in the preparation of this book, but makes no expressed or implied warranty of any kind and assumes no responsibility for any errors or omissions. No liability is assumed for incidental or consequential damages in connection with or arising out of information contained in this book. The Publisher shall not be liable for any special, consequential, or exemplary damages resulting, in whole or in part, from the readers' use of, or reliance upon, this material.

This publication is designed to provide accurate and authoritative information with regard to the subject matter covered herein. It is sold with the clear understanding that the Publisher is not engaged in rendering legal or any other professional services. If legal or any other expert assistance is required, the services of a competent person should be sought. FROM A DECLARATION OF PARTICIPANTS JOINTLY ADOPTED BY A COMMITTEE OF THE AMERICAN BAR ASSOCIATION AND A COMMITTEE OF PUBLISHERS.

LIBRARY OF CONGRESS CATALOGING-IN-PUBLICATION DATA

Mallett, William. Surface transportation congestion / William J. Mallett.
p. cm.
ISBN 978-1-60456-885-1 (softcover)
1. Traffic congestion--United States--Prevention. 2. Transportation--United States--Planning. 3. Traffic engineering--United States. I. Title.
HE355.3.C64M35 2008
338.3'1420973--dc22   2008025057

Published by Nova Science Publishers, Inc. ≃ New York

Contents

Preface	vii
Summary	1
Issues for Congress	7
Brief History of Transportation Congestion	19
Legislative History of Transportation Congestion	25
Transportation Congestion: Concepts, Measures, and Trends	33
The Costs of Transportation Congestion	53
Transportation Congestion Remedies	57
Concluding Observations	65
References	67
Index	81

PREFACE

Surface transportation congestion most likely will be a major issue for Congress as it considers reauthorization of the Safe, Accountable, Flexible, Efficient Transportation Equity Act — A Legacy for Users (SAFETEA), P.L. 109-59, which is set to expire on September 30, 2009. By many accounts, congestion on the nation's road and railroad networks, at seaports and airports, and on some major transit systems is a significant problem for many transportation users, especially commuters, freight shippers, and carriers. Indeed, some observers believe congestion has already reached crisis proportions. Others are less worried, believing congestion to be a minor impediment to mobility, the by-product of prosperity and accessibility in economically vibrant places, or the unfortunate consequence of over reliance on cars and trucks that causes more important problems such as air pollution and urban sprawl. Trends underlying the demand for freight and passenger travel — population and economic growth, the urban and regional distribution of homes and businesses, and international trade — suggest that pressures on the transportation system are likely to grow substantially over the next 30 years.

Although transportation congestion continues to grow and intensify, the problem is still geographically concentrated in major metropolitan areas, at international trade gateways, and on some intercity trade routes. Because of this geographical concentration, most places and people in America are not directly affected by transportation congestion. Consequently, in recent federal law, Congress, for themost part, has allowed states and localities to decide the relative importance of congestion mitigation vis-a-vis other transportation priorities. This has been accompanied by a sizeable boost in funding for public transit and a more moderate boost in funding for traffic reduction measures as part of a patchwork of relatively modest federally directed congestion programs. Congress may decide to continue with funding flexibility in its reauthorizationof the surface transportation programs. States and localities that suffer major transportation congestion would be free to devote federal and local resources to congestion mitigation if they wish. Similarly, congestion-free locales would be able to focus on other transportation-related problems, such as connectivity, system access, safety, and economic development. Alternatively, Congress may want to more clearly establish congestion abatement as a national policy objective, given its economic development impact, and take a less flexible and, in other ways, more aggressive approach to congestion mitigation.

Three basic elements that Congress may consider are (1) the overall level of transportation spending, (2) the prioritization of transportation spending, and (3) congestion pricing and other alternative ways to ration transportation resources with limited government spending. Congress also may want to consider the advantages and disadvantages of specific transportation congestion remedies. Hence, this book discusses the three basic types of congestion remedies proposed by engineers and planners: adding new capacity, operating the existing capacity more efficiently, and managing demand.

SUMMARY*

Surface transportation congestion most likely will be a major issue for Congress as it considers reauthorization of the Safe, Accountable, Flexible, Efficient Transportation Equity Act — A Legacy for Users (SAFETEA), P.L. 109-59, which is set to expire on September 30, 2009. By many accounts, congestion on the nation's road and railroad networks, at seaports and airports, and on some major transit systems is a significant problem for many transportation users, especially commuters, freight shippers, and carriers. Indeed, some observers believe congestion has already reached crisis proportions. Others are less worried, believing congestion to be a minor impediment to mobility, the by-product of prosperity and accessibility in economically vibrant places, or the unfortunate consequence of over reliance on cars and trucks that causes more important problems such as air pollution and urban sprawl. Trends underlying the demand for freight and passenger travel — population and economic growth, the urban and regional distribution of homes and businesses, and international trade — suggest that pressures on the transportation system are likely to grow substantially over the next 30 years.

Although transportation congestion continues to grow and intensify, the problem is still geographically concentrated in major metropolitan areas, at international trade gateways, and on some intercity trade routes. Because of this geographical concentration, most places and people in America are not directly affected by transportation congestion. Consequently, in recent federal law, Congress, for the most part, has allowed states and localities to decide the relative importance of congestion mitigation vis-a-vis other

---

* This report is excerpted from CRS report #RL33995, Updated February 6, 2008

transportation priorities. This has been accompanied by a sizeable boost in funding for public transit and a more moderate boost in funding for traffic reduction measures as part of a patchwork of relatively modest federally directed congestion programs.

Congress may decide to continue with funding flexibility in its reauthorization of the surface transportation programs. States and localities that suffer major transportation congestion would be free to devote federal and local resources to congestion mitigation if they wish. Similarly, congestion-free locales would be able to focus on other transportation-related problems, such as connectivity, system access, safety, and economic development. Alternatively, Congress may want to more clearly establish congestion abatement as a national policy objective, given its economic development impact, and take a less flexible and, in other ways, more aggressive approach to congestion mitigation. Three basic elements that Congress may consider are (1) the overall level of transportation spending, (2) the prioritization of transportation spending, and (3) congestion pricing and other alternative ways to ration transportation resources with limited government spending.

Congress also may want to consider the advantages and disadvantages of specific transportation congestion remedies. Hence, this report discusses the three basic types of congestion remedies proposed by engineers and planners: adding new capacity, operating the existing capacity more efficiently, and managing demand.

Transportation congestion most likely will be a major issue for Congress as it considers reauthorization of the Safe, Accountable, Flexible, Efficient Transportation Equity Act — A Legacy for Users (SAFETEA), P.L. 109-59, which is set to expire on September 30, 2009. By many accounts, congestion on the nation's road and railroad networks, at seaports and airports, and on some major transit systems is a significant problem for many transportation users, especially commuters, freight shippers, and carriers. Moreover, trends underlying the demand for freight and passenger travel — population and economic growth, the urban and regional distribution of homes and businesses, and international trade — suggest that pressures on the transportation system are likely to grow in the years ahead.

A number of experts and organizations believe that congestion has reached crisis proportions. In announcing a new National Congestion Strategy in May 2006, then Secretary of Transportation Norman Mineta stated that "congestion is one of the single largest threats to our economic prosperity and way of life."[1] In a similar vein, the Transportation Research Board (TRB) currently has congestion on its "critical issues" list as one of the most pressing problems of the transportation system, arguing "if the 20th century can be called the era of building, the 21st may be called the era of congestion."[2] More recently, in January 2007, the U.S. Government Accountability Office (GAO), for the first time, placed transportation financing and capacity on its list of high-risk federal programs and operations.[3]

Not everyone agrees that congestion is a major, national problem. Some see it as a minor impediment to mobility, others as an unfortunate by-product of prosperity and accessibility in economically vibrant places. Several environmental groups argue that congestion is less the problem than the over reliance on the cars and trucks that cause it. Indeed, this over reliance on highway transportation, they believe, leads to more important problems, such as suburban sprawl and air pollution. Furthermore, because the problem is geographically concentrated, most places and people in America do not suffer noticeable levels of congestion. Thus, many might question to what extent transportation congestion is a national problem warranting a federal government response. In uncongested regions, transportation problems are more often to do with basic connectivity of the transportation system, system access, and economic development.

Connectivity, system access, economic development, and congestion relief are some of the objectives of national transportation policy that also include mitigating the negative effects of transportation, such as deaths, injuries, and environmental damage. According to 49 U.S.C. § 101,

> The national objectives of general welfare, economic growth and stability, and security of the United States require the development of transportation policies and programs that contribute to providing fast, safe, efficient, and convenient transportation at the lowest cost consistent with those and other national objectives, including the efficient use and conservation of the resources of the United States.

To accomplish these objectives, the federal government regulates transportation activities and provides funding to encourage states and local governments to build and operate transportation infrastructure. Since the beginnings of this "federal-aid" system, there have been major debates about how these funds should be distributed and spent. An underlying tension throughout these debates has been whether to distribute funds to encourage the pursuit of nationally defined transportation goals, such as the building of the Interstate system, or to distribute funds equally between the states (according to a predefined formula) and allow them to pursue their own objectives.[4] In SAFETEA, about 90% of highway funds are authorized to be distributed by formula, and states are guaranteed by FY2008- FY2009 a 92% return on money paid into the highway account of the Highway Trust Fund.[5]

Because transportation congestion is geographically concentrated, Congress has tended to favor a state and local approach to solving transportation congestion in the recent history of the federal surface transportation program. This has been accompanied by several sizeable boosts in funding for public transit and traffic reduction measures directed to major metropolitan areas in an attempt to curb the negative effects of cars and trucks, including road traffic congestion. Congress also has enacted a patchwork of other programs to deal with congestion at the national level, with some success, but these have generally been relatively modest efforts. Consequently, the flexibility provisions of recent federal laws, and with them the equity provisions that attempt to return to each state the taxes paid by its highway users into the highway account of the Highway Trust Fund, have largely left it to the states, and in some cases metropolitan planning organizations, to decide funding priorities.

The extent to which Congress decides congestion is a national problem to be solved by federal dictates, and funding may be a major issue in reauthorization. Congress may decide its current "bottom-up" approach to planning and programming transportation improvements, with some modifications, is the best approach to congestion in the broader scheme of transportation priorities. Conversely, Congress may decide that congestion warrants a stronger role for the federal government. Three broad elements of the issue are discussed here: overall levels of transportation spending, the prioritization of transportation spending, and congestion pricing and other alternative rationing schemes that require limited government spending.

Although congestion is being experienced throughout the transportation system, including at ports and airports, this report is limited to a discussion of congestion associated with the surface transportation system — highways, public transit, and freight and passenger rail. Because these modes connect with ports and airports, there is some discussion of intermodal issues at these nodes as well, but the report does not discuss congestion in the waterway or airway systems per se. This report begins by outlining in broad terms some of the issues that Congress may face in the reauthorization debate. This is followed by a brief history of transportation congestion in the United States, and how Congress has dealt with the issue in the recent past. It then goes on to discuss transportation congestion concepts, measures, and trends, followed by information on the national costs of congestion. The final section lays out some of the major types of congestion remedies that have been proposed by transportation engineers, planners, and policy makers.

ISSUES FOR CONGRESS

Most experts agree that surface transportation congestion has grown over the past few decades and, moreover, that the demand for surface transportation services is likely to continue growing over the next few decades. According to one national assessment of highway congestion by the Texas Transportation Institute (TTI), total delay in 437 urban areas increased five-fold between 1982 and 2005, and delay per peak-period traveler almost tripled.[6] Anecdotal evidence suggests that overcrowding is a growing problem in some major transit systems and that conflicts between freight and passenger rail trains (commuter and intercity) are an issue for both. In the freight rail industry, the Congressional Budget Office (CBO) notes that average speeds, one indicator of congestion, are lower now than at anytime since the early 1980s except for the 1997-1998 period following the merger of Union Pacific and Southern Pacific.[7] With dramatic increases in foreign trade, many fear that ports and border crossings have become significant bottlenecks to the flow of commerce.

Despite these trends, the question remains as to whether or not congestion is a national problem and, therefore, should be a specific goal of national transportation policy. Although congestion has intensified and spread, congestion is geographically concentrated in major metropolitan areas, at international trade gateways, and on some intercity trade routes. Because of this geographical concentration, most states and localities do not suffer any appreciable transportation congestion directly. Moreover, some argue that even in places with relatively intense congestion problems, it only adds a few extra minutes to daily travel and that many actually enjoy the extra time alone in the car away from the pressures of work and family.[8] Seen in terms of an entire trip, including the time it takes to park and walk to the office, one expert believes the extra time caused by freeway delay is relatively minor.[9] Some even go so far as to suggest that much like a crowded restaurant or nightclub, congestion is a sign of success and its costs must be balanced against the benefits of access to jobs, stores, recreational amenities, etc. that congested regions provide.[10] Environmental organizations generally argue that road traffic congestion results from an unbalanced transportation system, one that favors cars and trucks, and that urban sprawl, air pollution, and noise, not road traffic congestion per se, should be the focus of national policy.[11]

The alternative view is that transportation congestion is a major problem, national in scope, and, if unchecked, a problem that will intensify and spread over the next 25 years. Many experts point out that although congestion may be highly localized, because transportation is a network that serves the U.S. population in a variety of ways, its economic effects are national. Most obviously, freight movement is largely dependent on a national transportation network in which a bottleneck in one place, such as southern California, may affect businesses and consumers in largely congestion-free Nebraska. Moreover, these experts point out the national network effects are becoming increasingly important as supply chains lengthen and become more complex. Similarly, although passenger transportation is mostly a local affair, congestion on roads that service airports and other passenger terminals may also result in inefficient intercity passenger travel, dragging down the productivity of businesses that rely on it for managing far-flung operations.

Local congestion may also be thought of as a national issue in that the places where it is found tend to be the hubs of the national economy and its costs, therefore, are not inconsequential in terms of the national economy. For instance, the 28 metropolitan areas that experienced 40 hours or more of annual delay per peak-period traveler (as measured in 2005 by TTI) account for more than 45% of total personal income in the United States (in 2005).[12] Most businesses rely, to one degree or another, on the efficient transportation of people locally, whether it is the transportation of managers to business meetings, workers to work, or customers to places where products are consumed. Research has shown that metropolitan areas with the largest labor markets tend to have the highest productivity.[13] Consequently, when added together, the local costs of congestion, some argue, are significant in national terms.

Another commonly expressed view is that given current trends in the supply and demand for transportation the problems of congestion will affect more people and more businesses in the future. Road traffic congestion, for instance, is growing fastest in the smaller urban areas included in the TTI study, though admittedly from a small base. However, research by the Federal Highway Administration (FHWA) shows a wider problem when it projects future demand on the current highway system.[14] Underlying these trends are broader trends in population and the economy. For example, the population is expected to reach 364 million by 2030, an increase of about 20% from 2007.[15] Over the same period, the CBO projects GDP to increase by about 70% (in real terms).[16] Furthermore, the FHWA predicts that freight movements will nearly double between 2002 and 2035.[17]

The federal surface transportation program approach to congestion tends to view it as a state and local issue, not as a major national problem. At least as far back as passage of the Intermodal Surface Transportation Efficiency Act (ISTEA) of 1991 (P.L. 102-240), Congress has tended to leave to the discretion of the states, within certain planning parameters, the relative weight to be placed on congestion mitigation vis-a-vis other transportation priorities. In this regard, many argue that governments and other stakeholders closest to transportation problems are in the best position to craft solutions. Another issue since the 1980s, with the near completion of the Interstate system, has been the controversy regarding state payments to and from the Highway Trust Fund (HTF), known as the "donor-donee" debate.[18] This debate focuses on the perceived fairness of the relative size of each state's payments to and receipts from the highway account of the Highway Trust Fund. Increasingly over the years, federal law has attempted to equalize these amounts rather than concentrate funding where needs are greatest. Several new federal programs to tackle congestion nationally have been developed, but, in dollar terms, these have been relatively modest.

Because state and local funding flexibility has been a significant feature of federal transportation policy since ISTEA, Congress may decide to continue with this approach in reauthorization. States and localities that suffer major transportation congestion would be free to devote federal and local resources to congestion mitigation if they wish. Similarly, congestion-free locales would be able to focus on other transportation-related problems, such as connectivity, system access, safety, and economic development. Alternatively, Congress may want to take a less flexible and, in other ways, more aggressive approach to congestion mitigation. Three basic elements to the problem that Congress may want to consider are (1) the overall level of transportation spending, (2) the prioritization of transportation spending, and (3) congestion pricing and other alternative ways to ration transportation resources.[19]

TRANSPORTATION SPENDING LEVELS

The amount of federal funding for surface transportation programs is a major issue during all reauthorization debates and will undoubtedly be an issue in the reauthorization of SAFETEA. Some observers contend that America is underinvesting in transportation infrastructure, resulting in deteriorating conditions and worsening performance, including growing congestion.[20] One alternative to addressing transportation congestion, in this view, is a significant increase in the overall level of infrastructure investment to deal with the existing backlog of projects and future needs. The most recent needs assessment by the U.S. Department of Transportation (USDOT) suggests that the cost to maintain the current condition and operational performance of the highway system is about 12% more annually than is being currently spent by all levels of government. For transit, the figure is 25%. Spending to improve conditions and reduce congestion would be greater than this.[21] It should be pointed out that, as with any attempt to estimate current and future system conditions and performance, there are a host of simplifying assumptions, omissions, and data problems that influence the results. Nevertheless, this analysis suggests that if total government spending is not increased above current levels, the physical condition of system elements may decline and congestion, particularly highway congestion, will continue to increase.

An alternative view of the overall level of government transportation spending is that it has not been dramatically deficient. In this view, deteriorating performance, and in some places deteriorating conditions, are the result of resources not being directed to the parts of the system that are in greatest demand and, therefore, have the greatest needs for maintenance and expansion. Indeed, USDOT's own analysis of historic spending patterns shows that total government spending in highways and transit, including capital spending, has generally kept pace with usage since the early 1980s, although the federal share has declined. Capital spending by all levels of government on highways per vehicle mile has remained relatively constant since about 1980, at around 2.5 cents per vehicle mile (in real terms).[22] Over this period, the federal share declined from close to 60% to a little under 40% at the end of the 1990s, but has since rebounded to about 44% in 2004.[23]

In terms of the nation's transit systems, the USDOT analysis shows that total government spending on capital and operations grew by approximately 80% between 1980 and 2004 (in real terms), much faster than passenger trips, which grew by 12%.[24] The federal share of total spending declined from 42% to 25% over this period.[25] The federal share of capital spending in 2004 was 39%, somewhat lower than the approximately 50% share that existed in the mid-1990s.[26] In 2004, the federal government funded about $36 billion of highway and transit capital expenditure, with 86% going to highways and 14% to transit.[27] The transit share increases to about 16% if all government spending is included.[28]

Consequently, assessments of highways nationally reveal that conditions have generally improved overall during the past decade, particularly in rural areas, but have declined in large urban areas.[29] Similarly, bridge conditions have improved, but to a much greater extent in rural areas than in urban areas.[30] As noted above, operational performance on the urban highway system has generally declined, but there are also growing pressures on the higher elements of the rural highway system, especially rural interstates.[31] Transit conditions and performance have remained about the same over the past decade, but rail system performance has declined to some extent.[32]

Some experts, however, believe that investment in the freight rail industry fell behind demand at some point over the past decade or so, leading to rail congestion and higher prices for shippers.[33] Freight rail, as a predominantly private industry, depends on investment received mostly from railroad profits or from money borrowed in capital markets to be paid back with future revenues. One view is that these sources of investment will be adequate to cope with future demand. Another view is that because of the great risks inherent in investing in rail infrastructure and the demands of shareholders, the railroads themselves will not be able to supply the necessary capital to expand capacity. In that case, some contend that government financial assistance will be needed, otherwise rail congestion will grow and more freight will be diverted to the roads.[34]

As the case of the railroads reminds us, not all transportation infrastructure investment comes from federal, state, and local government. The private sector is a major source of investment and not just in rail transportation. A flurry of recent major privatization efforts, such as the Chicago Skyway and the Indiana East-West Toll Road, have increased interest in this approach. Thus, some argue that there is a need for much greater investment in transportation, but that the federal government should consider using its resources to leverage private investment through public-private partnerships. Others argue that these types of public-private partnerships will be limited to only a few places with the highest profit potential and that investment could be quickly cut off if macroeconomic conditions change.

TRANSPORTATION SPENDING PRIORITIES

With growing pressure on transportation infrastructure but competing claims on governmental resources, another issue for congressional consideration is improving the efficiency of federal investments. Some argue that prioritizing investments may be a better way to deal with congestion mitigation than the scattershot, "more-is-better" approach. Several aspects of prioritizing federal transportation spending to mitigate transportation congestion could be of interest to Congress. These are prioritizing projects by location and project type, and the issue of mode-neutrality. Inherent in these discussions, of course, is how project decisions are made and the ways in which the relationships between federal, state, and local governments affect the outcome. This is another aspect of prioritization that may be of interest to Congress.

Continued federal transportation funding likely will be needed to maintain and operate the transportation system as a whole and to meet other national transportation goals such as rural access, urban mobility, safety, and national security. However, it can be argued that if mitigating congestion in the name of enhancing national mobility and economic productivity is a national goal, then federal funding will need to be focused in the places that promise the greatest return: those with the most congestion. The three major locales of transportation congestion are major metropolitan areas, some intercity trade routes, and foreign trade gateways.

An oft-cited argument for targeting federal resources toward congested places is that while the project costs of congestion mitigation are local, the benefits, at least in part, are regional or national in scope. In addition, fixing transportation bottlenecks is very often a hugely expensive proposition and, therefore, beyond the means of a single locality or state. Moreover, many point out that in addition to the pecuniary costs of large transportation facilities, costs associated with local environmental and social disruptions must be mitigated.

Another aspect of prioritizing federal funding to mitigate congestion is the way in which projects are planned and funded *within* states and regions. For the most part, project development and funding decisions are made by state departments of transportation (DOTs). Metropolitan planning organizations (MPOs) have assumed a greater role over the years, but not enough to fundamentally change the traditional federal-state intergovernmental relationship that has existed since the beginning of the Federal-Aid Highway Program.[35] One effect of this, some have suggested, is that highway funding tends to be funneled disproportionately toward rural areas at the expense of urban and suburban areas where needs, including congestion mitigation needs, are greatest. A study of Ohio found this to be the case because many municipal roads are ineligible for state funding, state gas taxes are limited by state law to highway projects, and state apportionments are made equally to counties without regard to needs such as population, miles of road, and traffic volumes.[36] Some of the same processes may also occur within metropolitan regions that comprise many local jurisdictions. For instance, some observers contend that local government officials are often more concerned about receiving their "fair share" of funding than they are about solving regional problems such as transportation congestion. MPOs also tend in most instances to be dominated by suburban areas at the expense of center cities because voting power is often not weighted by population size.[37] Of course, weighted voting is no guarantee that a central city will not be dominated by surrounding jurisdictions when collectively they comprise a larger share of the regional population.

A number of other factors have also been found to affect transportation investment decisions.[38] Broad stakeholder involvement requirements in federal law and, in some cases, the need for local voter approval can have a major influence on which types of projects move forward and which do not. For example, freight interests, a relatively minor constituency, argue that such requirements often lead to the prioritization of passenger projects over freight projects. In addition, state and local officials, needing to forge consensus on major investment decisions, tend to favor system preservation, maintenance, and operations projects because they are comparatively easy and quick to implement. By contrast, major capacity expansion projects are typically controversial and can take a decade or two to complete. Added to this is the fact that densely populated urban areas often have limited space available for major new infrastructure and that old and inadequate infrastructure can be very difficult and expensive to expand.

Choosing among the types of strategies that provide the most cost-effective reductions in congestion could be done in a number of ways. The most effective projects are likely to vary from place to place and situation to situation, requiring local solutions rather than national dictates. However, Congress may require project alternatives to be chosen after an assessment of the full benefits and costs, with congestion mitigation and economic efficiency as high priorities.[39] A major study of transportation in the United Kingdom found that projects aimed at relieving congestion "offer remarkably high returns, with benefits four times in excess of costs on many schemes, even once environmental costs have been factored into the assessment."[40] A different approach is a performance-based assessment in which a federal standard or goal is set, such as a certain level of congestion reduction, freeing state and local governments to determine the most efficient way of meeting the goal.[41]

Another important issue with respect to prioritization is "mode neutrality." Traditionally, federal surface transportation funding has been focused on highways and transit. This has made it difficult to fund projects involving modes that fall outside these categories, such as freight rail or multi-modal projects.[42] Program changes have been made over the years to allow greater flexibility, but some argue that these changes have not gone far enough. An opposing view is that when private transportation infrastructure providers are involved, it is very difficult if not impossible to properly assess the public benefits and costs of public subsidies. Others fear that subsidizing private businesses may substitute public investment for private investment with no net gain for the transportation system, or that such assistance may provide some businesses an unfair advantage over others.

Mode neutrality in transportation congestion mitigation is still an issue in the relative balance between funding highways and transit. Some argue that highway congestion cannot be solved by building more highway capacity or otherwise improving service because this only encourages or "induces" more people to travel by highway, thereby restoring the same, or an even higher, level of congestion. Instead, they contend that alternatives such as public transit in concert with land use measures to encourage the use of alternative modes of travel are the only way around congestion.[43] Others argue that so few people use transit to get to work, and even fewer for other reasons, that major new investments in transit capacity, except in a limited number of situations, are not likely to reduce highway congestion appreciably, if at all.[44]

The problem and empirical measurement of induced demand are a central element in many of the debates about road traffic congestion. The theory of induced demand suggests that building more road capacity will not solve road traffic congestion because it merely "induces" travelers using other modes, driving on other routes, or driving at other times of the day to travel on the new facility during the peak period, resulting in congestion as bad as that suffered before the expansion.[45] Some suggest it is even possible for congestion to become worse in the long run after a road is built or expanded because the new capacity encourages more development, resulting in proportionally more drivers than the new capacity added.[46] Attaining a definitive answer to this question is difficult because of the confounding factors of regional trends in population and employment growth and other things that lead to changes in transportation habits.[47] However, several studies show that although induced demand is real, it typically takes a number of years for the new capacity to be absorbed, suggesting that new capacity can reduce congestion in the medium term.[48] Moreover, other experts note that while congestion may reassert itself after the addition of major new capacity, the new facilities still serve more travelers than before even if service quality is poor, and the increase in travelers on the new or larger facility may take pressure off other facilities, improving travel over the whole network.[49]

CONGESTION PRICING AND OTHER ALTERNATIVE WAYS TO RATION RESOURCES

Many economists argue that transportation congestion is caused by the way in which service is rationed. In highway transportation, for example, because the marginal cost of driving is so low, congestion is the main method for rationing peak-period roadway space. Peak-period roadway space is in great demand for deep-seated reasons that have to do with the need for face-to-face interaction in economic and social situations. Thus, at certain times and in certain places, demand for roadway space exceeds supply and vehicles have to queue for the next available space to open up. It is argued that road traffic congestion could be reduced by using different rationing methods. One approach is to limit roadway space to certain types of vehicles or vehicles carrying a certain number of passengers, such as buses or high-occupancy vehicle (HOV) lanes. Another method is to ban a vehicle or driver from driving at certain times for one or more days a week. The method generally favored by economists, however, is to use some sort of pricing mechanism, known as *congestion pricing* or *value pricing*. Its supporters argue that not only does road pricing have the potential for solving congestion, it also promotes the most efficient use of highway infrastructure.

Detractors argue that road pricing unfairly favors higher-income drivers, may cause severe mobility problems where no reasonable alternative exists, and may, if it raises the cost of traveling in the most dense urban areas, lead to more sprawl and highway congestion farther out from the urban core. Another argument against tolling in general, of which congestion pricing is one form, is that drivers have often already paid for the infrastructure and its maintenance through taxes and fees, and so it amounts to a form of double taxation. Consequently, some suggest that such strategies should be used only to fund and manage new capacity or should not be used at all.

Demand for transit service in large cities is typically more concentrated, both in time and by direction, than demand for highway travel. The result can be vehicle overcrowding, service denial, and, because overcrowding tends to increase vehicle dwell times (i.e., time spent at a station or bus stop to discharge and pick-up passengers), overall slower speeds. Despite this, most transit agencies do not differentiate fares on the basis of peak/off-peak service but instead have flat-fare structures and offer unlimited ride passes.[50] As is often pointed out, higher peak-period fares would help to cover the higher costs of providing peak-period service and might persuade some travelers to travel during less busy periods. Even where higher peak-period fares are employed, however, they are not usually high enough to substantially reduce demand peaking. Proposals to introduce differentiated fare schemes to reduce overcrowding — or in places that have them to raise fares even higher at congested times or places — are often viewed skeptically as a way for a transit agency to generate more revenue, particularly from transit-dependent travelers. Others fear such schemes might push public transit users to drive instead, causing greater highway congestion.

In freight rail transportation, prices (or "rates" as they are more commonly known) are already the main mechanism used to manage supply and demand. Rates reflect the cost of providing freight rail service and demand. Demand for rail service is largely a function of the overall strength of the economy and the ability of rail transportation to compete with other modes, particularly trucks and barges. With strong demand and constrained supply, economic theory would suggest, all else equal, that rates will increase, providing greater resources for investing in expanding supply. Although the situation is complex, because not all else is equal, the evidence suggests that with greatly improved productivity and strong demand, the financial health of the railroad industry has improved substantially since deregulation. This has allowed railroad companies to make significant investments to maintain the current system and to increase capacity in some places.[51] Nevertheless, there is widespread concern that the railroads will not be able to make sufficient investments to keep up with demand.[52]

A number of reasons have been posited for the inability of railroads to invest sufficiently in new capacity to keep up with demand. Clearly, expanding capacity is a slow process, meaning it may take decades for supply and demand to find an equilibrium, if it ever does. Moreover, in many congested urban areas, railroads find it difficult to acquire land for new capacity.[53] Port areas that could benefit from new rail lines and terminal facilities are notoriously space-constrained. The railroads argue that they suffer several inequities that hinder their ability to finance new capacity. The railroads note that, unlike trucking and barge firms, they provide their own infrastructure and must bear the long-term risks associated with owning fixed assets. Furthermore, they argue, other modes pay less in taxes and fees than their use of public infrastructure would warrant, putting the railroads at a competitive disadvantage. Railroads also argue that they are subject to several industry-specific laws that raise their costs in comparison with their competitors. These laws include the Railroad Unemployment Insurance System and some remnants of the Interstate Commerce Act.[54] Ultimately, the railroads argue that despite improvements in their financial situation since deregulation, they continue to have problems earning enough to cover the cost of capital, hindering their ability to compete for financing in capital markets.[55]

In this context, a number of public policy alternatives have been suggested to alter the current rationing of public and private resources. One controversial proposal is to impose greater taxes and fees on truck and barge companies to "level the playing field" with railroads. Another is to provide government assistance to railroads to mitigate some of the risks they face, with the goal of increasing the level of investment and accelerating its current pace.[56] On the other hand, some contend that the railroads ought to make a greater financial commitment to solving problems where they impose high external costs, such as places where rail operations contribute significantly to highway congestion. For example, in 2002, northeastern Illinois was estimated to have about 1,700 highway-rail grade crossings that caused nearly 11,000 hours of motorist delay on a typical weekday.[57] Contributions by the railroads to highway-rail grade crossing improvements, such as grade separation projects, however, tend to be a relatively small share of the overall cost.

A final consideration in the rationing of resources is what might be called the costs of debate, review, and approval. Some argue that the costs of complying with federal, state, and local regulation stemming from the multitude of planning, environmental, and community involvement laws have substantially increased project costs since the 1960s. These costs include the direct compliance costs of staff time and the indirect costs of project delay that results in foregone opportunities in terms of improved mobility, safety, and the like. Most agree that these laws serve an important purpose and have several benefits. Nevertheless, many would like to reduce the delay caused by the unnecessary duplication of effort and coordination problems among the different parties.[58]

BRIEF HISTORY OF TRANSPORTATION CONGESTION

HIGHWAY TRANSPORTATION

In the early years of the century, before the mass production of motor vehicles, congestion generally referred to overcrowded trolley lines and trolley cars in major cities and downtown streets filled with pedestrians and horse-drawn passenger and goods vehicles.[59] For most of the 20th century, however, transportation congestion meant road traffic congestion. The rapid rise of motor vehicle ownership, particularly with the introduction of Ford's Model T in 1908, together with rudimentary road and traffic control systems, made urban road traffic congestion a major transportation problem by the 1920s.[60] Federal, state, and local governments responded with a significant road-building effort in this period, although road traffic congestion was largely "solved" by the Great Depression and the Second World War.

During the Second World War, with the massive diversion of resources to the war effort, automobile use was widely discouraged. Public transit ridership boomed again during this period, reaching an all-time high in the United States in 1946 of 23.4 billion trips.[61] However, car ownership and motor vehicle travel rose rapidly after the war causing another bout of concern with road traffic congestion, particularly in and around cities.[62] Congestion and the threat of future congestion were among the reasons cited by President Eisenhower in his push to create the Interstate Highway Program,[63] although he was against the idea of urban interstates, preferring instead bypasses that would allow through traffic to avoid the central cities. Nevertheless, the cities themselves were insistent that urban interstates were needed to solve urban congestion problems, and Congress obliged in the Federal-Aid Highway Act of 1956 and the Highway Revenue Act of 1956 (P.L. 84-627).[64]

Road capacity expanded rapidly following the passage of the 1956 acts that also created the Highway Trust Fund. Less than 20 years later, by the end of 1974, about 36,000 miles of the 42,500 mile system were complete, with another 2,800 miles under construction.[65] Together with the improvement of other urban and rural road networks, road capacity (measured by paved centerline miles of highways and streets[66]) grew at about the same rate as motor vehicle travel from the mid-1940s to the mid-1960s (Figure 1). The problem of road traffic congestion never disappeared in major cities, but in the 1970s, the most vexing highway transportation problems were energy, air quality and other environmental issues, and highway safety.

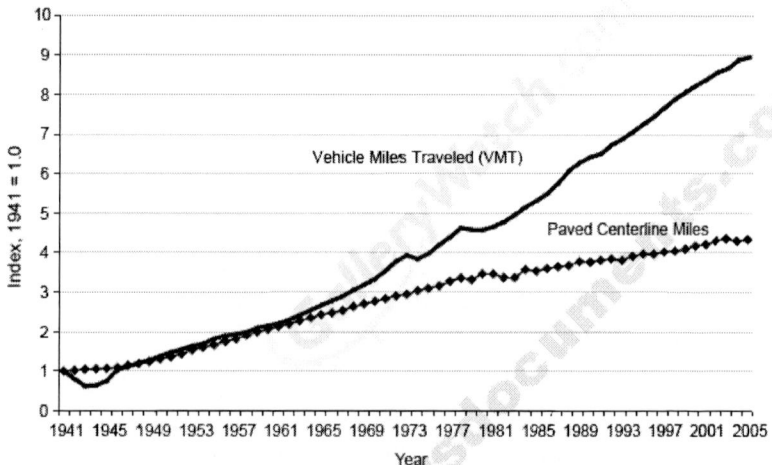

Source: U.S. Department of Transportation, Federal Highway Administration, *Highway Statistics* (Washington, DC, various years).

Note: Paved centerline miles is the length of roads paved with some type of bituminous, Portland cement concrete, or brick surface as measured along the center in one direction.

Figure 1. Motor Vehicle Travel and Road Capacity, 1941-2005.

The growth in road capacity and motor vehicle travel began to diverge in a major way during the 1970s, as shown in Figure 1. Except for slight dips associated with the oil shocks of 1974 and 1979, motor vehicle travel continued to grow apace. At the same time, growth in road capacity slowed as the interstate system neared completion, maintenance requirements began to absorb more resources, and building new capacity became more expensive and time-consuming as a result of new environmental laws. Consequently, road traffic congestion began to climb quickly again in the 1980s and has continued to rise ever since.

PUBLIC TRANSPORTATION

Public transportation congestion has not been a major issue since the end of the Second World War, when transit ridership was at an all-time high. On the contrary, the major issue, particularly through the 1950s and 1960s, was the overall lack of riders resulting from increases in motor vehicle ownership, suburbanization, and other changes in work and leisure.[67] By the early 1970s, transit ridership was only a quarter of what it had been at its peak in 1946, dropping from a high of 23.4 billion trips to a low of 6.5 billion trips. In response, many streetcar systems were abandoned in favor of diesel buses, and privately owned and operated transit systems were taken over by public authorities. Public transportation has undergone something of a resurgence since the mid-1970s with the building of a number of new rail systems, particularly light rail, but also heavy rail and commuter rail. Since then, transit ridership has increased modestly to about 10 billion trips in 2005.[68] To put this in context, however, the proportion of all trips made on transit declined by half between 1969 and 2001, as trips by other modes, particularly in personal motor vehicles, increased to a much greater extent.[69]

Although not as widespread as road traffic congestion, peak-period transit overcrowding has become an issue in some cities with large numbers of transit commuters and heavily congested roads and railways, such as New York; Chicago; San Francisco; Washington, DC; and Boston. Peak-period overcrowding on the subway in Washington, DC, for instance, has led to proposals for substantially higher fares at the most heavily used times and stations.[70] In addition, because most transit buses do not run on roads with controlled access (e.g., high-occupancy vehicle [HOV] and bus lanes), road traffic congestion also affects bus riders.

Freight Transportation

Until relatively recently, congestion has not been a major issue in freight transportation. The building of the interstates, together with the existing rail, water, and pipeline systems, provided adequate surface freight capacity from the 1960s through the 1980s. According to many analysts, the biggest problem at this time was antiquated federal regulation from laws dating to the late 19th and early 20th centuries. Administered mainly by the now defunct Interstate Commerce Commission (ICC), these regulations controlled prices and competition, leading to some major inefficiencies in the transportation of goods. Deregulation beginning in the late 1970s sparked a major reorganization within and across modes that overall has provided shippers with cheaper, more efficient freight transportation and much greater choice.

In railroading, for instance, federal regulation made it difficult to abandon little-used or unprofitable lines. Thus, although railroad mileage peaked as early as 1916, it changed little for the next 60 years. Overcapacity was a significant contributor to the financial difficulties of the railroads that reached crisis proportions in the 1970s and subsequently led to deregulation of the industry through the Staggers Rail Act of 1980 (P.L. 96-448). The Staggers Act made it much easier for major railroads to abandon lines or to sell or lease them to non-Class I railroads.[71] Since then, the miles of track owned and operated by Class I railroads have dropped precipitously from 271,000 in 1980 to 162,000 in 2006.[72] Non-Class I railroad mileage consequently has grown, although modestly. Despite less track, railroads today are able to move more freight because technological changes allow them to run heavier, longer, and faster trains. Indeed, freight rail ton-miles increased by 93% between 1980 and 2006.[73] This has also been accomplished with relatively fewer locomotives, freight cars, and employees, marking huge productivity gains since deregulation.

Deregulation also played a major role in the reorganization and growth of the trucking industry. New laws such as the Motor Carrier Act of 1980 (P.L. 96-296) and other changes freed up trucking companies to more directly compete against each other, allowed the entry of new firms, and encouraged the development of efficient truck operation and routing. The results have been generally lower prices and higher-quality and more reliable service.

Among other things, deregulation played an important role in the shift toward what has been called "coordinated logistics," defined as "the integration of distinct logistics activities, such as cross-modal coordination or the bundling of transportation and inventory control."[74] Deregulation helped remove many of the modal and jurisdictional barriers between carriers. Moreover, with industry consolidation and improvements in productivity and profitability, carriers were able to introduce new technologies and develop innovative services. For instance, over the past few decades, trucking and railroad companies have created networks of trailer-on-flatcar service that combine the advantages of rail and truck transportation. With cheaper and more timely deliveries of goods, shippers have been able to save production and distribution costs by developing longer and more complex supply chains and by cutting back on their inventories of goods. Coordinated logistics, therefore, has raised the importance of transportation in the logistics process and has placed greater emphasis on seamless networks of multiple transportation modes.

Coordinated logistics has also been spurred on by extraordinary growth in foreign trade. Foreign trade as a percentage of U.S. Gross Domestic Product (GDP) has grown from 11% in 1970 to 26% in 2005.[75] Consequently, the amount of goods moving through foreign trade gateways — ports, border crossings, and airports — has skyrocketed. For instance, waterborne merchandise trade almost tripled between 1970 and 2006, from 581 to 1,565 million tons.[76] This growth has placed great pressure on the gateways themselves, but also on the transportation networks that serve them — primarily roads and rail lines — and the connection between modes. Among other problems, most of these gateways are located in large urban centers that suffer from high levels of road traffic congestion and have limited space for facility expansion. Many experts now believe the efficiency gains resulting from deregulation and other changes have largely run their course.[77] After declining for years, the cost of logistics to U.S. businesses appears to be increasing, partly because of congestion (see Figure 2). In railroading, many lines and terminals are running at or near full capacity. With little or no slack in the system, railroads have become more susceptible to disruptive incidents, such as late loadings and unloadings, breakdowns, and poor weather. Another problem as rail lines reach capacity is the growing conflict between freight and passenger trains (Amtrak and commuter) that, for the most part, use the same lines. As a result, delays are multiplying for both freight and passenger trains, particularly in major urban areas that generate a lot of freight and passenger traffic. In trucking, productivity is now largely dependent on road congestion, the supply of qualified truck drivers, and fuel costs.

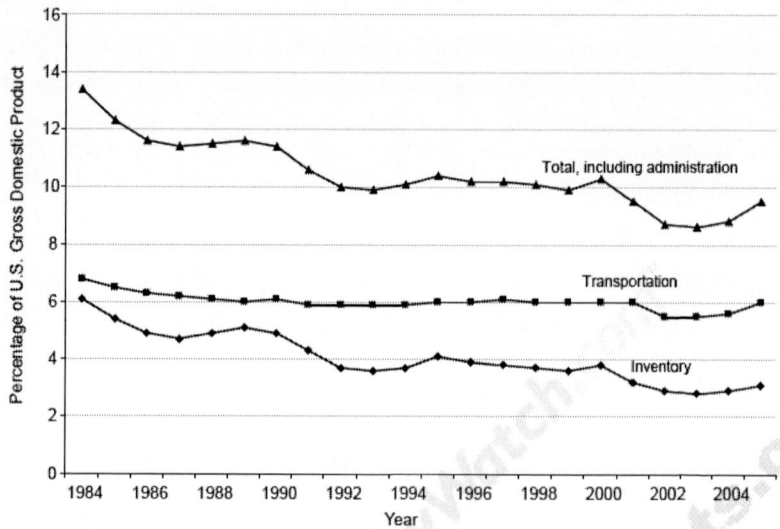

Source: Council of Logistics Management, State of Logistics Report (Washington, DC, 2006).

Figure 2. U.S. Cost of Logistics, 1984-2005 (percentage of Gross Domestic Product).

LEGISLATIVE HISTORY OF TRANSPORTATION CONGESTION

In line with the rapid growth of motor vehicle ownership and travel, federal surface transportation policy for most of the 20th century focused on road connectivity and capacity, particularly with a view to providing basic access in rural areas and then intercounty and interstate roads. Urban road traffic congestion warranted a certain amount of attention in the early Federal-Aid Highway Acts, including the Federal-Aid Highway Act of 1956. Federal transit funding, beginning in the 1960s, was also partly predicated on the argument that it would relieve road traffic congestion.[78] As the interstate building program neared completion in the 1980s and road traffic congestion was growing apace, federal policy makers began a fundamental reassessment of surface transportation policy. The result was the Intermodal Surface Transportation Efficiency Act of 1991 (ISTEA), P.L. 102-240.

INTERMODAL SURFACE TRANSPORTATION EFFICIENCY ACT OF 1991 (P.L. 102-240)

In the deliberations of the congressional committees that culminated in the passage of Intermodal Surface Transportation Efficiency Act (ISTEA), there was recognition that urban road traffic congestion was a major problem.[79] Unlike in the past, however, some viewed road capacity building as a flawed strategy for dealing with the issue. This view was summed up by Senator Daniel Moynihan in the introductory statement of the

Senate report. Talking about the building of the interstate system, he argued the following:

> [T]he plain fact is that traffic congestion has grown during this period of massive highway construction. We have to face the fact that even if we had greater resources than we do, adding to highway capacity does not any longer seem a promising road to increased highway efficiency.[80]

Congressional leaders also expressed the concern that solutions to transportation problems that encouraged more driving would lead to more air pollution, thereby undermining the provisions of the recently enacted Clean Air Act Amendments of 1990 (CAAA), P.L. 101-549. A third major concern was that scarce resources should be used first and foremost to maintain and improve the current highway system over system expansion.

Rather than design a new road-building program, leaders in both the House and the Senate sought to fashion a program to enhance the *efficiency* of a transportation system that was largely in place. This new program would be based on highway system maintenance; more transit funding; greater funding flexibility; intermodalism; enhanced state and metropolitan planning; improved operations, including development and deployment of advanced technologies (e.g., technologies to improve roadway monitoring, enhance traveler information, and enable the electronic payment of tolls); and efforts to improve safety, energy efficiency, and pollution control. The new surface transportation bill also required the designation of a new National Highway System (NHS) to prioritize federal help for the most heavily traveled routes of the Interstate Highway System, the Strategic Highway Network, and Federal-Aid Primary System.[81]

A fundamental theme in the development of ISTEA was that states and localities should be free to fashion their own solutions to local problems, a tenet that became known as "flexibility." While recognizing that congestion was a problem, the committees understood that it was not a problem everywhere, hence the need for flexibility. In this regard, the House Public Works and Transportation Committee report noted, "The new system reflects the Committee's recognition of the need to relieve congestion in urban and suburban America, while at the same time addressing the mobility and access needs of Rural America."[82]

In reworking the surface programs, the large Surface Transportation Program (STP), authorized at $24 billion over the life of the bill, was at the core of the flexibility provisions. STP funds were made available for highway capital projects but could be "flexed" to transit if desired and if certain other conditions were met. As the House Committee noted, "For those with congested urban areas, flexibility may mean more transit solutions, while for rural areas or those experiencing economic growth, flexibility may mean more highways."[83] Within certain parameters, flexibility was also provided for switching funds between the different parts of the highway system, as projects could be on any part of the system except local and rural minor collectors. Moreover, STP funds could be used for a bridge project on any public road, not just those on the federal-aid system.

ISTEA also authorized a substantial increase in federal transit funding over previous authorizations, nearly $32 billion over the life of the bill. Many expected that additional funding would be devoted to transit from flexed STP funds and another new program, the $6 billion Congestion Mitigation and Air Quality program (CMAQ), which was authorized to provide new funds for projects to help states and localities meet the requirements of the Clean Air Act Amendments (CAAA) of 1990. Funding was aimed primarily at reducing pollutants emitted by reducing motor vehicle travel, particularly single-occupant vehicle travel. Because the most polluted places tend to have the worst road traffic congestion, it was believed that many projects funded under CMAQ to reduce pollution would reduce road traffic congestion as well. However, CMAQ prohibited spending on more traditional congestion relief projects, such as new road capacity that would be primarily used by single-occupant drivers. In addition, building new capacity could violate the requirements in the Clean Air Act and ISTEA that state and metropolitan plans "conform" to the emissions levels set forth in the air quality State Implementation Plan (SIP) as required by CAAA. A 10-year assessment of the program found that about 44% of CMAQ funds were spent on transit projects and another 33% on traffic flow improvement projects such as incident management, HOV lanes, and traffic signal improvements.[84]

ISTEA also advanced a few other congestion-related programs that were federal program innovations. First was the idea of intermodalism — planning and financing projects that enhance the links between modes. In this regard, states and metropolitan areas were required to consider the transportation systems as whole in the planning process and to include participation from all stakeholders, including the freight community. Funds were also made available for highway projects to accommodate other transportation modes and for carpool projects, such as fringe and corridor parking facilities and programs, and bicycle transportation and pedestrian walkways. Second, ISTEA placed more emphasis on funding highway operations, including the establishment of a new program to fund the development and deployment of advanced technology in transportation, known as the Intelligent Vehicle/Highway Systems Program (IVHS). Now known as Intelligent Transportation Systems (ITS), the program was originally authorized with $660 million over the six year life of the act. Third, to enhance the ability of metropolitan areas to coordinate and fund the development of their transportation systems, ISTEA increased the responsibilities of metropolitan planning organization (MPOs) and required the development of congestion management systems at both the metropolitan and state level. The requirement for a congestion management system at the state level was subsequently dropped in the National Highway System Designation Act of 1995 (P.L. 104-59).[85] Fourth, ISTEA provided funding for up to five projects in the Congestion Pricing Pilot Program and allowed greater use of federal funds on toll roads than in the past.

NATIONAL HIGHWAY SYSTEM DESIGNATION ACT OF 1995 (P.L. 104-59)

ISTEA required the designation of a new category of highways, the National Highway System (NHS), to be worked out in consultations between the USDOT and the states. The designation of the 155,000-mile NHS system was the primary purpose of the NHS Act. However, the NHS Act included several other provisions amending the federal programs, some with relevance to the issue of mobility and congestion. Among them were the authorization of two new financing mechanisms: the State Infrastructure Bank (SIB) pilot program and what became known as Grant Anticipation Revenue Vehicle (GARVEE) bonds. The SIB pilot project allowed a handful of states to use some of their highway and transit funds to capitalize a revolving fund. The GARVEE bonds were developed from Section 311 of the NHS Act that expanded the use of federal-aid highway funds for bond financing. A number of intermodal projects, including the Alameda Corridor project, were advanced because of these new provisions.

TRANSPORTATION EQUITY ACT FOR THE 21ST CENTURY (P.L. 105-178; P.L. 105-206)

The Transportation Equity Act for the 21st Century (TEA-21), as amended (P.L. 105-178; P.L. 105-206), enacted June 9, 1998, maintained the essential structure of the programs created in ISTEA with an increase in funding (in nominal terms) of 40%. Of the total $218 billion authorized, $177 billion was allocated for highways and $41 billion for transit, although TEA-21 continued and enhanced the flexing of monies between modes as introduced by ISTEA in 1991.[86]

Several programs begun in ISTEA were retained and expanded under TEA-21. CMAQ was retained with more funding ($8.1 billion) and expanded eligibility criteria. ITS funding was raised to $1.282 billion, and a new ITS program, the Commercial Vehicle Information Systems and Networks (CVISN) Program, was established and funded at $184 million. With the ultimate goal of improving the efficiency and safety of commercial motor vehicle operations, the CVISN program was created to make use of information systems and communications networks by developing industry standards and demonstrating potential benefits. Three areas were initially targeted under the CVISN program: safety information exchange, credentials administration, and electronic screening.[87] The Congestion Pricing Pilot Program was renamed the Value Pricing Pilot Program and funded at a higher, though still very modest, level ($51 million).

TEA-21 also created a few new programs. Some of these came under the banner of innovative financing, including the Transportation Infrastructure Finance and Innovation Act (TIFIA) and the Railroad Rehabilitation and Improvement Financing (RRIF) program. TIFIA was to provide up to $10.6 billion in credit assistance to large projects of national significance (generally projects over $100 million). The RRIF program was set up to provide loan and loan guarantees up to $3.5 billion, of which not less than $1 billion was to be available to non-Class I railroads. Two new infrastructure grant programs — the National Corridor Planning and Development Program and the Coordinated Border and Infrastructure Program — were also created and jointly funded at $140 million per year for FY1999 through FY2003.[88] The first was conceived primarily as an economic development tool (although congestion costs were one factor to be used in determining projects) and the second was intended to alleviate congestion and improve mobility at the borders. Since FY2000, nearly all the funds in this program have been earmarked in appropriation bills.

SAFE, ACCOUNTABLE, FLEXIBLE, EFFICIENT TRANSPORTATION EQUITY ACT — A LEGACY FOR USERS (P.L. 109-59)

After a number of hearings prior to reauthorization of TEA-21 in which transportation congestion was a major focus, the initial legislative proposal from the House of Representatives (H.R. 3550) in the 108th Congress included a number of new provisions in Subtitle B, entitled "Congestion Relief." Two provisions were seen as being particularly innovative. The first was the Motor Vehicle Congestion Relief Program, which would require states with an urbanized area over 200,000 to set aside apportioned funds under several existing programs to be spent on projects that enhance capacity and relieve congestion. The proposed set-aside was 10% of a state's total apportionments multiplied by the percentage of the state's population in urbanized areas of 200,000 or more. The second innovative proposal was to fund ITS technologies at a much higher level and to speed up their deployment. H.R. 3550 would have authorized about $4 billion during FY2004-FY2009, with about $3 billion of this amount for expedited deployment. This was up from about $230 million per year toward the end of TEA-21 (not including federal-aid highway funds allocated by the states to deploy ITS).[89]

H.R. 3550 proposed a new $6.6 billion allocated program called Projects of National and Regional Significance to fund important high-cost facilities ($500 million or more or greater than 75% of a state's annual apportionment), including freight rail projects eligible under Title 23 U.S.C. Also included in the bill was a new Freight Intermodal Connectors program to be funded by formula at the level of $1.37 billion over six years and a Freight Intermodal Distribution Pilot Grant Program funded at $30 million over five years as a takedown from the Freight Intermodal Connectors authorization.[90] This latter program was intended to provide grants to facilitate intermodal freight transportation initiatives at the state and local levels to relieve congestion and improve safety, and to provide capital funding to address infrastructure and freight distribution needs at inland ports and intermodal freight facilities. As passed by the House, two tolling provisions were also included in H.R. 3550, one to permit states to allow drivers to pay to use HOV facilities as part of a variable toll-pricing program and the other to permit the construction of new lanes on interstates to be funded by tolls.[91]

The reauthorization of the surface transportation programs was not passed in the 108th Congress but was eventually completed in the 109th Congress and signed into law by the President on August 10, 2005. The Safe, Accountable, Flexible, Efficient Transportation Equity Act — A Legacy for Users (SAFETEA) provides a general increase in transportation funding with a six-year total of $286.4 billion for programs from FY2004 through FY2009. This represents a 31% increase in nominal terms over the $218 billion provided over the six years of TEA-21 (FY1998-FY2003).[92]

Table 1. SAFETEA Authorization Levels, by Legislative Titles and Selected Programs, FY2005-FY2009 (in millions of dollars)

Selected SAFETEA Title/Program	Total Authorization FY2005-FY2009
Title I — Federal Aid Highways	**199,490.476**
Interstate Maintenance Program	25,201.595
National Highway System	30,541.833
Bridge Program	21,607.422
Surface Transportation Program	32,549.757
Congestion Mitigation & Air Quality Improvement Program (CMAQ)	8,609.100
National Corridor Infrastructure Improvement Program	1,948.000
Coordinated Border Infrastructure Program	833.000
Projects of National & Regional Significance	1,779.000
National Corridor Planning & Development & Coordinated Border Infrastructure Programs	140.000
Freight Intermodal Distribution Pilot Grant Program	30.000
Value Pricing Pilot Program	**59.000**
Title II — Highway Safety	**3,131.592**
Title III — Public Transportation	**45,313.000**
Title IV — Motor Carrier Safety	**2,519.829**
Titles V-X (excluding rescission of unobligated balances of highway contract authority in Title X)	**5,003.940**

Source: CRS Report RL33119, Safe, Accountable, Flexible, Efficient Transportation Equity Act —A Legacy for Users: Selected Major Provisions, coordinated by John W. Fisher.

As enacted, SAFETEA largely retains the structure of the surface transportation programs begun under ISTEA, with a large proportion of funding going to the established "core" highway programs (such as the Surface Transportation Program, the National Highway System, the Interstate Maintenance Program, and the Bridge Program) and public transportation.[93] The Congestion Relief subtitle of SAFETEA contains just one program, the new Real-Time System Management Information Program. This program, with no separate funds of its own, is designed to encourage states to develop a real-time traffic information system to improve highway operations and reduce congestion. The rest of the Congestion Relief programs, as proposed in H.R. 3550, were either shifted elsewhere in the act or deleted. ITS funding was not retained as a separate program but was "mainstreamed" as an eligible category in the core programs. CMAQ continues at a higher funding level, and project eligibility is expanded to include projects that might have a more direct impact on congestion. Table 1 shows the authorization levels of SAFETEA's titles and some selected programs for FY2005 through FY2009.

SAFETEA does provide states with slightly more latitude in using tolls to build or expand interstate capacity and to improve operational efficiency to reduce congestion. The Value Pricing Pilot Program was reauthorized at a higher level: $11 million for FY2005 and $12 million annually for FY2006-FY2009. In addition, SAFETEA includes provisions for a limited number of pilot projects to test the viability of the use of tolling on existing facilities including HOV facilities and for tolling to fund new interstate capacity.

SAFETEA also created the new Projects of National or Regional Significance program, but with funding set at $1.779 billion for FY2005 through FY2009, not $6.6 billion as proposed in H.R. 3550, and all the funds earmarked in the act. The new Freight Intermodal Connectors program was dropped before final passage of the bill, but the Freight Intermodal Distribution Pilot Program remained with $30 million authorized through FY2009. Again, this $30 million was earmarked in the bill. SAFETEA also reauthorized the Coordinated Border Infrastructure Program as a new apportioned program, with funding set at $833 million from FY2005 though FY2009.

Existing innovative funding provisions were extended and modified to some degree in SAFETEA. For instance, the minimum project size for TIFIA projects was reduced from $100 million to $50 million for most projects and from $30 million to $15 million for ITS projects. SAFETEA also allowed for broadened use of SIBs and Private Activity bonds. The RRIF was expanded tenfold under SAFETEA, from $3.5 billion to $35 billion in loans. Of this, $7 billion is reserved for non-Class I railroads. The legislation also added to the list of priorities in using such loans "enhancing rail infrastructure capacity and alleviating rail bottlenecks." SAFETEA also added a new federal grant program for relocating rail track that interferes with motor vehicle traffic.

TRANSPORTATION CONGESTION: CONCEPTS, MEASURES, AND TRENDS

Transportation congestion exists when demand for a transportation facility or vehicle is greater than its capacity and the excess demand causes a significant drop in service quality, such as speed, cost, and comfort, depending on the mode and specific situation. For example, when too many drivers compete for road space, the result is usually a significant drop in traffic speed but also higher vehicle operating costs and, with bumper-to-bumper, stop-and-go conditions, an increase in driver stress. In freight railroad transportation, train speeds may suffer when demand begins to reach capacity, and because shippers directly pay for access to rail infrastructure, higher rates theoretically may be another indicator of congestion. Depending on the situation, congestion in public transit may result in vehicle overcrowding — possibly resulting in service denial and reduced passenger comfort — slower vehicle speeds, and higher peak-period fares.

From the viewpoint of a multi-modal passenger trip or freight shipment, the possibility for congestion exists not only within each mode but also in the connections between modes. Poor or overstretched intermodal connections are another part of the transportation system that may damage service quality. Moreover, inefficient intermodal connections may cause problems within a mode as unexpected delays interfere with other trips and shipments farther down the line. For example, a delayed ship-to-truck transfer in a major metropolitan area may result in the truck traveling during peak-period traffic.

Ideally, transportation congestion should be defined and measured from the perspective of the end user — a traveler or a freight shipment.

Congestion, therefore, could be measured by the extent to which excess demand slows or otherwise harms a passenger trip or freight shipment from the origin to the destination.[94] In some situations, such as the transportation of packages by an express carrier, such as UPS and FedEx, it may be possible for the carrier to collect data and monitor movements for business purposes. However, in most situations, for public policy purposes, because measuring trips from origin to destination is difficult to accomplish in a large scale and meaningful way, measures of congestion typically focus on service problems within a mode. Moreover, within each mode, many measures of congestion are limited to a specific transportation facility. This is especially the case in highway transportation. For example, highway engineers typically refer to speed or level of service (LOS) on a particular road segment. Measurements on these segments are then sometimes aggregated to develop systemwide measures of highway congestion.

Mode-specific and facility-specific measures of congestion are not wholly satisfactory indicators of capacity problems in transportation service because they fail to measure aggregate impacts across the whole system. On the other hand, some transportation experts have noted that the focus on facility congestion instead of the effect of congestion on passenger and freight trips may also overstate its importance. For instance, freeway congestion may not be as bad as it seems if seen in the context of an entire automobile commute trip, including the time it takes to park and walk to the office.[95] Similarly, it might be true that the effect of freight bottlenecks might not be as bad as is generally believed if seen from the perspective of the entire supply chain.

Whether facility-based or trip-based, another criticism of transportation-based congestion measures is that they ignore the land-use context within which travel is taking place. In transportation planning parlance, they measure mobility but not accessibility. Accessibility explains the seeming paradox of why the most congested places are also the most economically vibrant, even when the congestion is long lived. Manhattan, for example, may be one of the most congested places on earth, but it also provides access to an enormous number of opportunities in terms of homes, jobs, retail outlets, restaurants, recreation, etc. A study of accessibility in Minneapolis, MN, for example, found that while traffic congestion more than doubled between 1990 and 2000 (measured in annual delay per person), access to opportunities by car, in this case the number of jobs, increased more quickly.[96] Seen from this perspective, the performance of the transportation system, in concert with land-use, actually improved in the 1990s rather than deteriorated, as congestion data alone would suggest.

Unfortunately, as it stands today, national data do not exist to examine the effects of congestion on accessibility as opposed to mobility. Nor do we have the means to examine the effects of congestion on passenger trips and freight shipments from end-to-end, including the efficiency of intermodal connections. The transportation congestion measures employed in most instances, including in this report, are both facility- and modally-based, with the inadequacies this entails. Several measures of congestion, particularly in freight rail and public transit, are gross indicators of capacity utilization using aggregate measures across the whole system. Moreover, no measures of intermodal terminal congestion per se exist. The measures of congestion presented here, nonetheless, represent the best available information today using publicly available data.

MEASURES AND TRENDS IN ROAD TRAFFIC CONGESTION

Efforts to define and measure road traffic congestion have increased over the past few decades as congestion itself has grown.[97] Still, congestion has proven difficult to measure at the national level because of the size and diversity of the highway system and because traffic problems can occur anywhere at any time of the day or night for a number of different reasons. Moreover, what constitutes a "congestion problem" is highly subjective. One frequently cited national road traffic research effort is the Urban Mobility Program at the Texas Transportation Institute (TTI). TTI defines traffic congestion as an excess of demand in relation to supply (or capacity) such that travel speeds are slower than normal, where normal is defined as free-flow speed. TTI derives travel speeds by relating the theoretical capacity of a roadway segment to the average number and type of vehicles traveling the segment. Speed estimates are then used to calculate travel delay. TTI uses data from FHWA's Highway Performance Monitoring System.[98]

Travel delay measures the extra time it takes to make a trip and can be expressed in several different ways, such as total delay, delay per traveler, and as a travel time index. The travel time index measures the ratio of travel time in the peak period to travel time at free-flow conditions. Thus, a Travel Time Index of 1.35 indicates a 20-minute free-flow trip takes 27 minutes in the peak-period.

In related research, TTI is developing measures of travel time reliability. Travel time reliability measures the variability of travel times. When the highway system is unreliable, travelers and shippers must build in extra time to avoid being late. TTI measures travel time reliability via its Buffer Time Index (BTI). The BTI measures the extra time needed to ensure that a traveler or freight shipment will arrive on time according to a predetermined standard, typically 95% of trips. A BTI of 43%, for instance, indicates that a traveler needs to add an extra 43% to the average travel time of a trip to arrive on time 19 out of 20 times (95% of trips).[99]

Some suggest that reliability is more important to both travelers and shippers than average delay. It seems reasonable to propose that most commuters would prefer to spend an extra 5 minutes to and from work each day than to endure an unexpected delay of 50 minutes on just one journey a week, a delay causing problems with arriving at work on time or picking up a child from school or daycare. Similarly, shippers often place greater value on being able to predict reliably when a shipment will arrive than on the speed with which it got there. In some cases, such as just-in-time manufacturing and distribution operations, shippers and carriers can face penalties for making late or, in some cases, early deliveries.

In its annual Urban Mobility Report, TTI aggregates road segment estimates for an entire urban area system of freeways and arterials. The same methodology has been used by other researchers to identify and measure delay and, in some cases, reliability at specific places, such as bottlenecks,[100] truck bottlenecks,[101] and border crossings,[102] as well as roads on the federally adopted National Highway System.[103] The FHWA is using similar measures to examine congestion on major travel corridors defined by Interstate routes, such as I-5 traversing California, Oregon, and Washington. However, in this research program, FHWA is using data collected from trucks themselves using Global Positioning System (GPS) technology.[104]

One of the main criticisms of TTI's work on urban road traffic congestion is that it does not directly measure congestion in any urban area, but relies instead on estimates of congestion based on a number of theoretical relationships. For a time, this meant that TTI was unable to account for improvements in speeds resulting from operational improvements — such as freeway entrance ramp metering, incident management programs, and traffic signal coordination programs — nor the effects of public transit. TTI has since begun including these variables in its models, but the overall criticism that its estimates of congestion are not direct empirical measurements still stands.

Another major criticism has to do with the estimation of congestion by comparing traffic speeds to free-flow conditions. A number of experts point out that such models can never fully account for induced traffic and that, as problematic as this may be theoretically, as a practical matter, eliminating congestion for all peak-period travelers is wholly unrealistic because the costs would be overwhelming. Thus, congestion-free peak-period travel in major metropolitan areas "is a purely notional idea, not a conceivable description of the world we might choose to provide for."[105] Moreover, using free-flow speed in the calculation of congestion can lead to some results that do not square with reality. For instance, if widening a road improves the peak-period average speed but is accompanied by a proportionally greater increase in the speed limit, the calculated amount of congestion will increase after the improvement. In addition, a small change in average conditions, such as a decrease of a few miles an hour, may appear to be a significant congestion problem when measured over a large number of drivers.[106]

Empirical research on the relationship between freeway speed and vehicle flow shows maximum vehicle throughput at something less than free-flow speed, about 50 miles an hour. This too brings into question a congestion calculation based on free-flow speed. As Figure 3 shows, when there are few vehicles traveling on a freeway segment, as might be the case very early in the morning, average speeds are high, at about 60 miles per hour (mph), but overall throughput is low, at around 300 vehicles per lane per hour. As volumes build, vehicle throughput increases to around 1,800 vehicles per lane per hour and average speeds decline by about 10 to 15 mph. At this point, as the number of vehicles coming onto the road continues to increase, the volume of vehicles begins to overwhelm capacity and speeds decline precipitously. As speeds decline in this instance, vehicle throughput declines.[107]

Overall, this line of criticism concludes that estimating congestion using the unattainable ideal of free-flow conditions, and with it the costs of congestion (see below), tends to overstate its impact on society. This and other criticisms notwithstanding, the TTI estimates of urban road traffic congestion are widely used because they provide the only national picture of road traffic congestion on an annual basis and, hence, are useful for monitoring changes in congestion over time. Nevertheless, figures purporting to quantify the billions of hours of time lost (and their associated monetary value), numbers often used in newspaper headlines to dramatize the problem, ought to be viewed somewhat skeptically.

A very important finding from the work by TTI and others is that both roadway demand and roadway capacity are subject to short-term and long-term variations. Demand varies by day of week, time of day, and season, and in response to planned special events, such as professional football games, music festivals, and the like. Most road traffic congestion occurs on weekday mornings and evenings because of trips associated with jobs and school. Roadway capacity, on the other hand, is defined by the type of facility (number of lanes, access, etc.), its condition, and by events that may temporarily reduce capacity, such as traffic incidents, work zones, weather, railroad crossings, toll facilities, and commercial truck pickup and delivery in urban areas.[108]

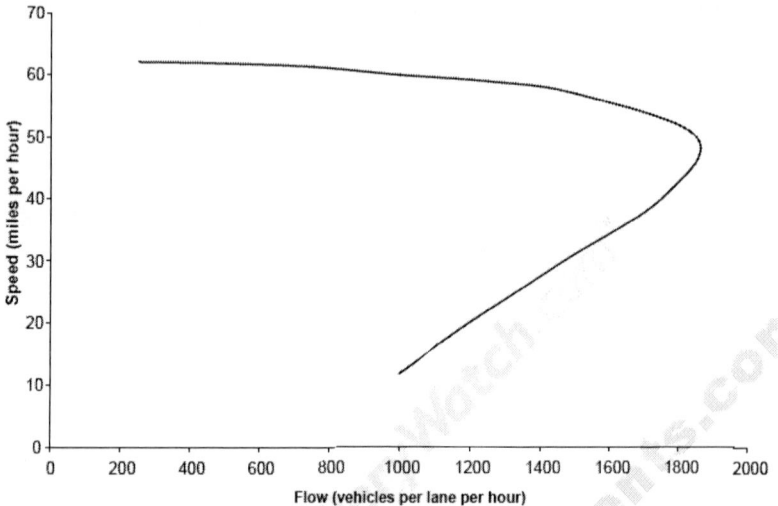

Source: Downs, Anthony, *Still Stuck in Traffic*, Brookings Institute Press (Washington, DC, 2006).

Figure 3. The Relationship Between Speed and Vehicle Flow on Freeways

According to the current research, about 40% of urban road traffic congestion is caused by capacity problems and another 5% is caused by poor signal timing (Figure 4). About 55% of congestion is the result of a temporary loss of capacity, with incidents (crashes, disabled vehicles, etc.) accounting for 25%, weather 15%, work zones 10%, and other events 5%.[109]

Current Trends in Road Traffic Congestion

Most experts agree that urban road traffic congestion has intensified and become more widespread during the past quarter century. TTI data from 437 urban areas covering the period 1982 through 2005 indicate that total travel delay has increased five-fold and delay per peak-period traveler has nearly tripled.[110] On average, delay increases with city size, but delay in small urban areas (those with a population of less than 500,000) has grown more quickly during this time period. Figure 5 demonstrates this in the 85 urban areas for which TTI provides detailed data. In addition, the morning and evening rush periods have lengthened and a greater share of roadways are congested. For instance, in the Louisville metropolitan area — a medium-sized urban area with a population of about 900,000 that covers parts of Kentucky and Indiana — the share of the road system congested has risen from 35% in 1982 to 52% in 2005. Moreover, the number of "rush hours" has increased from 4.2 hours per day to 7.2 hours.

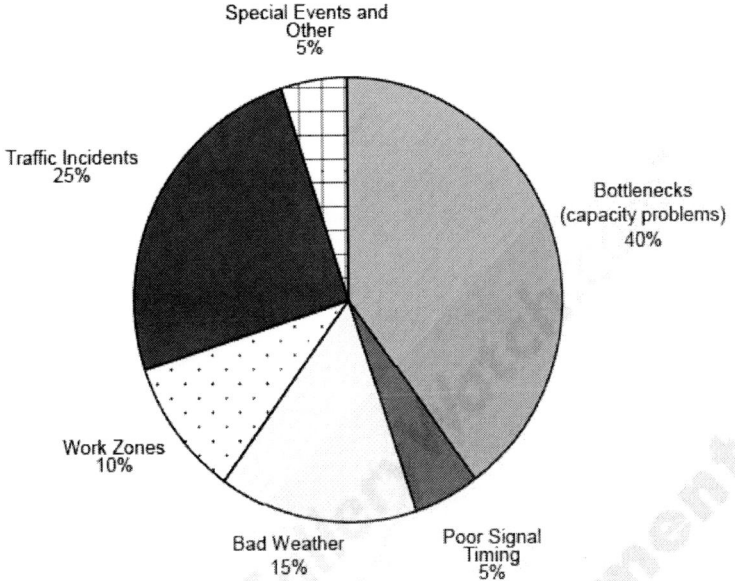

Source: Cambridge Systematic and Texas Transportation Institute, *Traffic Congestion and Reliability: Trends and Advanced Strategies for Congestion Mitigation*, report prepared for U.S. Department of Transportation, Federal Highway Administration (September 1, 2005).

Figure 4. Proximate Causes of Road Traffic Congestion.

Despite becoming more widespread, road traffic congestion is still heavily concentrated in a few of America's largest urban places. The 10 largest urban areas by population account for nearly one-half of total delay, though only about one-quarter of the U.S. population and the top 20 account for two-thirds of total delay and one-third of the population. Los Angeles suffered the most delay in 2005, with 72 hours of annual delay per peak-period traveler and a Travel Time Index of 1.5.

Urban road traffic congestion has increased because motor vehicle travel has grown rapidly, outstripping the existing road capacity and efforts to add new capacity and improve throughput with operational treatments. In the 437 urban areas studied by TTI, daily vehicle miles traveled on freeways grew by 128% between 1982 and 2005 and by 77% on arterials, while freeway and arterial lane-miles increased by only 41% and 37% respectively. Nationally, lane-miles grew by 4% and VMT by 87% during this period.[111]

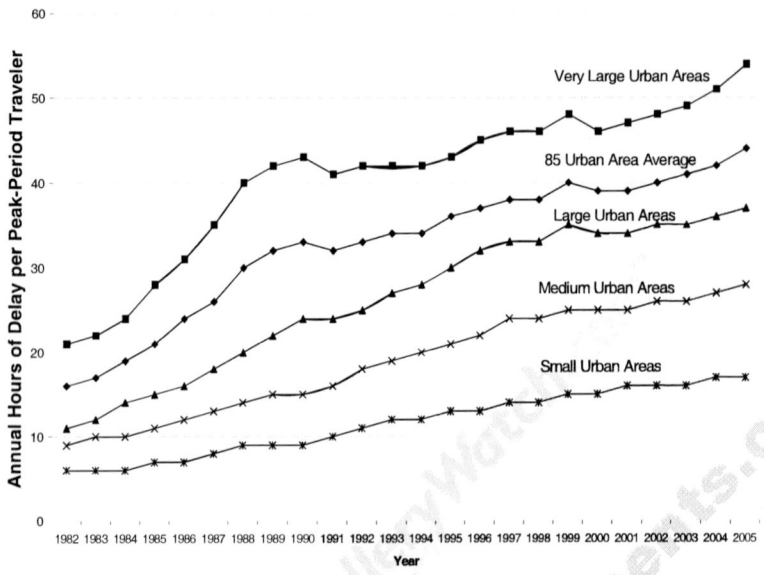

Source: Texas Transportation Institute, *Urban Mobility Report 2007* (College Station, Texas, 2007).

Figure 5. Road Traffic Congestion, 1982-2005.

Motor vehicle travel has grown rapidly for a number of reasons, including substantial growth in population, jobs, and national income; increased vehicle availability; and growth in metropolitan areas, particularly the suburbs. Between 1980 and 2005, the United States added 69 million people (a 30% increase), 42 million to the ranks of the employed (a 43% increase), 86 million motor vehicles (a 53% increase), and gross domestic product (GDP) grew by 113% in real terms.[112] Both population and job growth have been concentrated in metropolitan areas, most especially in low-density suburban rings that are difficult to serve with public transit. A metropolitan suburb-to-suburb commute is today, by far, the most common type of commute.[113] As result, most people drive alone to work — 77% in 2005, up from 64% in 1980. Over the same period, the share of commuters using transit hovered around 5%.[114]

These trends have been bolstered by an increase in the number and widespread availability of motor vehicles. The number of personal motor vehicles (cars, sport-utility vehicles, pickups, and minivans) per licensed driver passed 1.0 some years ago and continues to climb. In 2005, the average number of personal motor vehicles per driver was 1.16. That same year, only about 8% of households were without a vehicle.[115] The low price of gasoline has also contributed to enhancing the attractiveness of motor vehicles as a transportation option. For about 20 years beginning in the mid-1980s, the pump price of gasoline was below $2.00 per gallon (in 2006 dollars) in real terms, lower than at any time from 1918 on.[116]

Many of these same factors — population and income growth — together with economic complexity and globalization have led to more demand for commercial truck transportation. Since 1980, truck traffic has grown slightly faster than passenger traffic.[117] Although a lot of truck milage is made on long intercity trips, about half of truck VMT is made in urban areas, contributing significantly to urban traffic congestion, particularly near urban-based industrial facilities, ports, and border crossings.[118]

Many of the same factors generating vehicle travel and congestion are expected to continue growing. The Census Bureau expects the population to reach 364 million by 2030, an increase of about 20% from 2007.[119] Two-thirds of this population growth, and with it a significant portion of new road traffic, is expected to occur in just seven states: Florida, California, Texas, Arizona, North Carolina, Georgia, and Virginia. Over the same period, the CBO projects that GDP will increase by about 70% (in real terms).[120] FHWA's Highway Performance Monitoring System includes state-based estimates of future VMT growth.[121] The annual growth rate is projected to be 1.92%, with rural VMT growing somewhat faster than urban areas (2.15% average annual versus 1.79%).[122] The Freight Analysis Framework projects that freight tonnage by truck will double between 2002 and 2035.[123]

None of this is inevitable, and a few counter trends may slow the growth in VMT and peak-period travel. For example, although the age at which people are retiring from the workforce has begun to tick upwards over the past few years, baby boomers will begin retiring in large numbers in a few years. This may slow the growth in the number of workers. Some have suggested that as baby boomers age, they may begin to favor denser neighborhoods that are easier to serve with transit, thereby reducing the growth in VMT. Others believe there may be a reduction in work travel associated with flexible schedules, such as a compressed work week and telecommuting.

Interurban Road Traffic Congestion

Most, though not all, road traffic congestion is experienced in urban areas. An FHWA study of truck travel in freight-significant corridors — Interstate routes that span urban and rural areas — showed that a good deal of delay and reliability problems derive from the urban portion of trips.[124] Nevertheless, rural travel has grown faster than urban travel during the past 25 years. Between 1980 and 2005, rural VMT per lane mile grew by 65%, whereas urban VMT per lane mile grew 41%.[125] Estimates by FHWA of peak-period congestion on the federally adopted National Highway System in 2002 and a projection to 2035 suggest a much more widespread congestion problem. In 2002, FHWA's analysis of congestion found that it was largely confined to highway links in large urban areas. However, by 2035, assuming no change in physical road capacity or operational improvement, FHWA expects congestion to intensify in those areas and to spread to intercity corridors throughout the country.[126]

Road Bottlenecks

A number of studies have attempted to locate, characterize, and quantify bottlenecks in the highway system. TTI defines bottlenecks as "locations where the physical capacity is restricted, with flows from upstream sections (with higher capacities) being funneled into them."[127] One study found 233 major highway bottlenecks in 2002, defined as places with 700,000 hours of delay annually. This was a 40% increase in major bottlenecks from the 167 bottlenecks found in 1999. Of the 233 major bottlenecks in 2004, 24 had more than 10 million hours of delay in a year.[128] Freeway to freeway interchanges account for most bottleneck delay. According to another study, highway bottlenecks affecting large volumes of trucks accounted for 243 million hours of truck delay in 2004.[129] A third study on bottlenecks associated with summer vacation travel ranked the top 25 destinations likely to suffer the worst traffic delay in 2005.[130]

Road Congestion at International Gateways

Other potential bottlenecks in the transportation system are foreign trade gateways. Rapid growth in international trade over the past few decades has placed enormous pressure on these gateways — land border crossings, certain airports, and water ports — and the road and rail infrastructure that supports them. By value, in inflation-adjusted terms, international merchandise trade increased by 160% between 1980 and 2005.[131] Growth in value terms has been particularly rapid on the Mexican and Canadian borders and on the Pacific Coast, although the Atlantic Coast continues to handle the most trade (Figure 6). These trends are likely to continue with the growing globalization of production and consumption. Indeed, the FHWA expects foreign trade tonnage to more than double between 2002 and 2035.[132]

Although no comprehensive time-series data for congestion at land gateways nationwide exist, numerous studies have found delay and unreliable travel times at certain heavily used crossings. In 2004, daytime (8:00 a.m. to 6:00 p.m.) wait times for trucks entering the United States from Canada averaged 8.5 minutes, and those from Mexico averaged 7.3 minutes. However, daytime wait times at Laredo, TX, averaged nearly 21 minutes, and at Port Huron, MI, the average was 25 minutes.[133] Although they provide a basis of comparison, these averages mask the variability of delays that are probably more important. At land border crossings, congestion is caused by three main problems: inadequate transportation infrastructure to handle the volume of cars and trucks, import and security processing, and general urban road traffic congestion.[134] Some studies have suggested that border delay and reliability problems have more to with institutional and staff issues, such as inspection staffing levels at periods of high demand, than infrastructure problems, although this may depend on the specific crossing.[135] Similarly, delays at water ports may be caused by inadequate road and rail infrastructure, general road congestion, and customs and security requirements. Indeed, one of the big challenges at international gateways in the past few years has been balancing passenger and freight mobility with the need for heightened security in the wake of the terrorist attacks of 2001.[136]

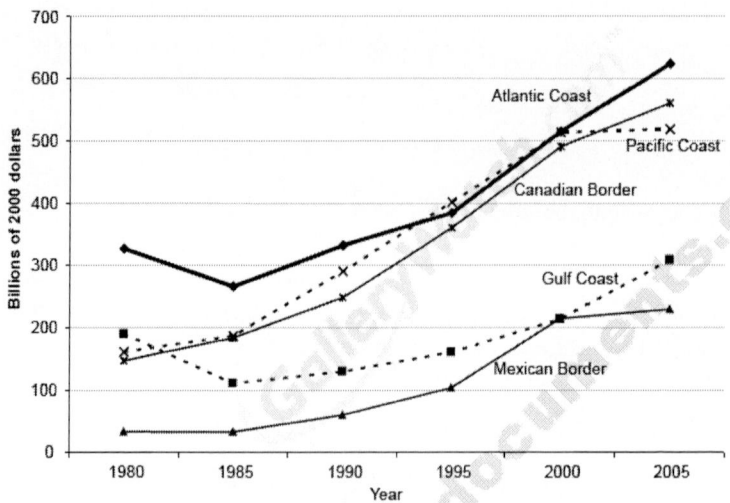

Source: U.S. Department of Transportation, Federal Highway Administration, *Freight Facts and Figures 2007* (Washington, DC, 2007).

Figure 6. U.S. Merchandise Trade by Region, 1980-2005.

MEASURES AND TRENDS OF CONGESTION IN PUBLIC TRANSIT

The main public transit modes in the United States — bus, commuter rail, heavy rail, and light rail — have different but overlapping characteristics that influence the causes and impacts of congestion. All public transit modes have the potential for vehicle overcrowding, but they differ in terms of system congestion. Transit buses typically run on roads in the general traffic stream and, therefore, are affected by road traffic congestion. In many cities, light rail systems have their own rights of way, but running at grade with limited separation can cause conflicts between rail and road traffic. Commuter rail service runs over rail lines that also carry freight and intercity passenger trains and, therefore, is subject to many of the same causes of delay and unreliability. Heavy rail (subway) systems have their own rights of way and, thus, are not subject to conflicts with other modes. However, subway system congestion is theoretically possible at peak periods when the number of trains running on the track begins to reach the design maximum, known as line capacity, and passenger loads affect station dwell times.[137] When running at full capacity, the lack of redundancy in the system also magnifies the effect of incidents such as a train breakdown.

Transit ridership grew 15% between 1980 and 2005. Over that time, bus ridership was virtually unchanged, while commuter rail and heavy rail grew by 51% and 33%, respectively. Light rail ridership almost tripled during these years because of the construction of several new systems.[138] Although all urban areas and many rural areas provide some sort of transit service, transit usage is heavily concentrated in a few large urban areas. Bus transit is widely provided, but only 34 metropolitan areas have one or more major forms of rail transit (defined here as commuter rail, heavy rail, and light rail). In 2004, 10 metropolitan areas accounted for 75% of all urban transit trips in the United States (see Table 2). The New York metropolitan area alone accounted for nearly 40% of all urban transit trips.

Table 2. Top 10 Metropolitan Areas by Transit Usage, 2004

Urbanized area	Rank	Annual Unlinked[a] Passenger Trips (thousands)	Cumulative % Urban Transit Trips	U.S. Pop.
New York, NY-NJ-CT	1	3,383,886	38	6
Los Angeles, CA	2	606,843	45	11
Chicago, IL-IN	3	582,786	52	14
Washington, DC-VA-MD	4	442,936	57	16
Boston, MA-NH-RI	5	396,087	61	17
Atlanta, GA	6	363,326	65	19
Philadelphia, PA-NJ-DE-MD	7	350,518	69	21
San Francisco-Oakland, CA	8	199,369	71	22
Seattle, WA	9	156,256	73	23
Miami, FL	10	151,222	75	25
United States, urban total		8,852,131		

Sources: U.S. Department of Transportation, Research and Innovative Technology Administration, Bureau of Transportation Statistics, *State Transportation Statistics 2006* (Washington, DC, 2007), table 4-3; U.S. Census Bureau, *Statistical Abstract of the United States, 2007* (Washington, DC, 2007), tables 17 and 25.

a. Unlinked passenger trips is the number of passengers boarding transit vehicles. A transit trip from origin to destination may involve one or more than one unlinked trips.

There are no direct measures of public transportation congestion available regularly on a national basis. Two indirect measures of congestion are average vehicle utilization, as a measure of vehicle overcrowding, and average operating speeds, as a measure of system congestion.[139] Vehicle utilization, as measured by the USDOT, is "calculated as the ratio of the total number of passenger miles traveled annually on each mode to total number of vehicles operated in maximum scheduled service in each mode, adjusted for the passenger-carrying capacity of the mode in relation to the average capacity of the Nation's motorbus fleet."[140] The USDOT notes that these two variables are related as "changes in the capacity utilization of rail vehicles have influenced these vehicles' operating speeds through changes in dwell times. As vehicles become more crowded, they take longer to unload and load, increasing wait at stations and hence passengers' total travel time."[141]

Average vehicle utilization data for urban transit systems show that passenger volumes in relation to service capacity are greatest on rail, particularly commuter rail. The higher level of commuter rail utilization is due to the longer average trip lengths with seating capacity only and to the limited time service is available. According to the FTA, utilization rates have generally declined since 2000/2001 (Figure 7). These data are bolstered by data on average speed that show little change in the average speed of non-rail modes, mainly buses, but a slight decline in speeds for rail transit. Non-rail speeds averaged 13.7 miles per hour in 1995 and 14.0 mph in 2004, but rail speeds declined from 26.6 to 25.0 mph over this period.[142]

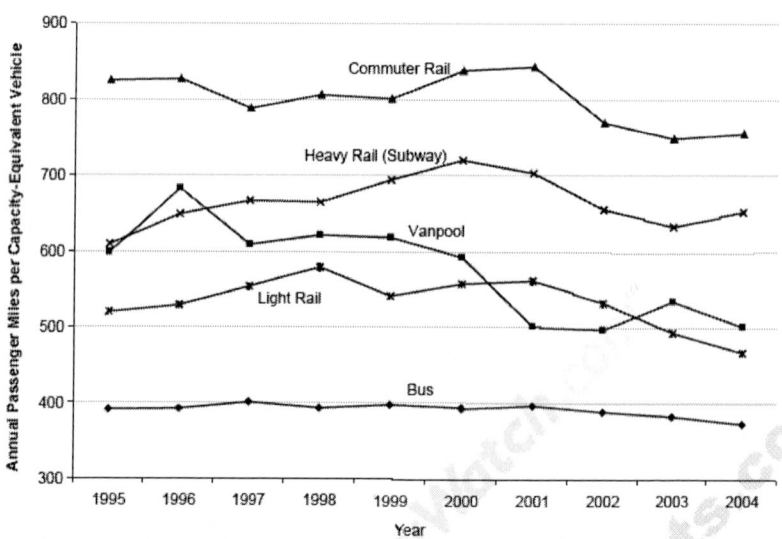

Source: U.S. Department of Transportation, Federal Administration and Federal Transit Administration, *2006 Status of the Nation's Highways, Bridges, and Transit: Conditions and Performance* (Washington, DC, 2007).

Figure 7. Transit Vehicle Utilization, 1995-2004.

Nevertheless, anecdotal evidence points to overcrowding problems on some rail transit systems, such as Washington's Metro and Boston's T. This suggests that these national average utilization data, which average over time and across place, may not fully capture rail transit overcrowding and system congestion in certain cities at certain times.

MEASURES AND TRENDS OF CONGESTION IN RAIL

Freight Rail Congestion Measures

The rail network is made up of a system of mainlines, spurs, sidings, yards, intermodal terminals, and places where the lines of different railroad companies come together (known as interchanges). Complexity is added by the physical characteristics of the thousands of tunnels, bridges, and overpasses with different clearances, the number and type of highway-rail grade crossings, and the thousands of miles of track with different load-bearing capacity and parallel lines. For the most part, this railroad infrastructure is owned and operated by private companies engaged in the transportation of freight. However, in some places, freight trains share space with passenger trains belonging to Amtrak and, in some urban areas, commuter rail operators.

In contrast to the way highway transportation works, decisions about accessing the rail system are controlled by a central authority — each railroad — that determines when a shipment will be transported and for what price. Thus, capacity problems tend to appear in a different form than they do on the highways and must be measured in different ways. Moreover, because the rail system is primarily private, the government has chosen not to collect and publicly disclose detailed data related to congestion. As a result, some indications of congestion problems are impressionistic and anecdotal.

In a free-market, when demand outstrips supply for a good or service, the price rises until an equilibrium between the two is found. One indicator of congestion in the rail industry, therefore, is freight rates. Unfortunately, understanding the relationship between capacity and prices is difficult as best. Rates are affected by any number of other variables, including the competition of other modes. Morever, rates can be regulated after the fact to protect "captive shippers." Capacity problems may also result in deterioration in service quality or no service at all. For example, in some cases, there may be a promise to transport a shipment at a certain price, but this shipment may be delayed as the operating railroad waits for space on the network. In other cases, some shipments may be denied access to the system completely and will have to travel by another means of transportation.

In theory, centrally controlled access to the rail system should avoid the queuing seen on highways; however, in practice, delay and unreliability do tend to increase as the number of trains on the system reaches maximum capacity. This derives from the complexity of determining the timing and routing of trains with different dimensions, such as single- or double-stacked containers, carrying different commodities over long distances, and the rules that must be followed to ensure that trains do not collide, particularly in places that are not signal-controlled. In addition, tight schedules can be upset by unforeseen incidents such as accidents, bad weather, and breakdowns and by interference with passenger trains that, by federal law, are supposed to have priority over freight trains.

Publicly available measures of freight rail congestion are traffic density, speed, and freight rates. None of these conclusively proves that congestion is a problem because they are all influenced by other things, such as efficiency gains derived from improved technology. Traffic density, as the Association of American Railroads (AAR) notes, "measures the average system-wide freight carrying utilization of the railroad track infrastructure. A higher figure indicates greater utilization efficiency, but can signal the risk of congestion."[143] Speed can be measured by average train speed or by net ton-miles per train hour (freight speed). Again, slower speeds might be an indication of a congestion problem, but they might also be related to other factors, such as the mix of commodities being transported and length of haul. Average cost is measured by freight revenue per ton-mile. TRB notes that this has been declining for years because of productivity growth, excess capacity, and deregulation. It notes a slowing of the rate of decline or even a pronounced increase might be indicative of a congestion problem.[144]

Trends in Freight Rail Congestion

The three measures of capacity utilization — traffic density, average freight speed, and freight rates — all suggest a growing congestion problem in the industry. This is supported by anecdotal evidence of trip times and bottlenecks. Since rail deregulation in 1980, Class I rail freight ton-miles have increased 93%, from 919 billion to 1,772 billion, while miles of track have decreased 40%. Traffic density measured by millions of revenue ton-miles per mile of track, therefore, has increased from 3.4 in 1980 to 10.9 in

2006 (Figure 8).[145] Moreover, these data exclude demands placed on the system by intercity and commuter passenger rail operations.

Source: Association of American Railroads, *Railroad Facts* (Washington, DC, various issues).

Figure 8. Freight Rail Traffic Density, 1980-2006.

The average speed of freight moved by rail, measured by net ton-miles per train hour, grew substantially in the 1980s but has since declined (Figure 9). Consequently, as CBO notes, the average speed is "now lower than it has been since the early 1980s, except for the turbulent 1997-1998 period following the merger of Union Pacific and Southern Pacific."[146] Another expert estimates that over the past 10 years, trip times have increased by about 25%-50% for general merchandise rail traffic.[147]

Average freight rates, measured by freight revenue per ton-mile, have declined substantially since deregulation from 5.3 cents per revenue ton-mile to 2.4 cents (in constant 2000 dollars). However, over the past decade the decline in rates slowed, and in the past few years rates have increased. Rates in 2006 were 14% higher in real terms than they were in 2003 (see Figure 10).[148] It is not clear, however, if this is indicative of a new upward trend in rates, nor is it clear how this relates to capacity problems in the industry.

Source: Association of American Railroads, *Railroad Facts* (Washington, DC, various issues).

Figure 9. Average Speed of Freight by Rail, 1980-2006.

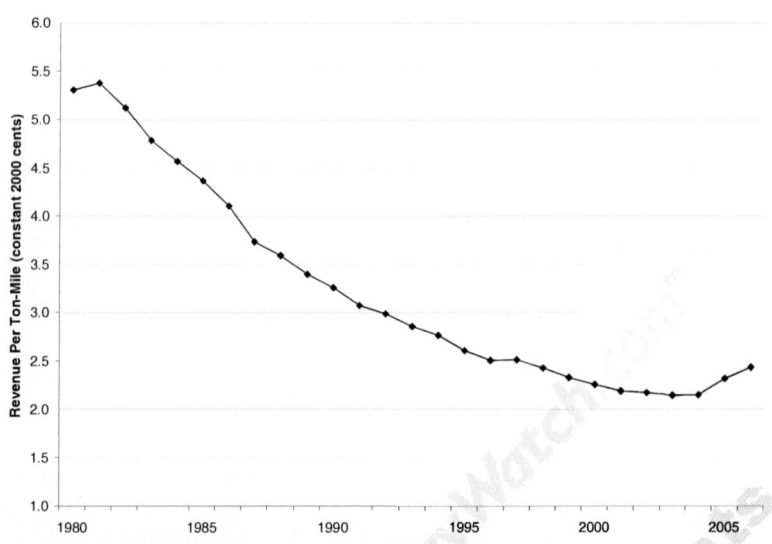

Sources: Association of American Railroads, *Railroad Facts* (Washington, DC, various issues); U.S. Bureau of Economic Analysis, "Gross Domestic Product."

Figure 10. Average Freight Rates, 1980-2006 (constant 2000 cents).

Like road traffic congestion, freight rail congestion is generally limited to a few key locations. Research completed for the Association of American Railroads indicates that about 3% of the freight rail network has demand at or above capacity, with another 9% near capacity. Some major bottlenecks include, among others, the network in and around Chicago, Kansas City, Atlanta, and Memphis as well as the rail corridors from San Francisco to Los Angeles and Los Angeles to Tucson, Arizona. In the Chicago region, congestion is compounded by the lack of connectivity between the several different railroads serving the area whose route systems are focused on states east and west of the Mississippi River.[149]

Intercity Passenger Rail (Amtrak) Congestion Measures

Congestion problems in intercity passenger train travel — trains operated by the National Railroad Passenger Corporation, known as Amtrak — are somewhat akin to those of the freight railroads discussed above. Except for the 500 miles it owns in the Northeast Corridor (NEC), intercity passenger trains operated by Amtrak run on rail lines that are owned and operated by freight railroads. As freight movements have grown, so too have the conflicts between freight and passenger trains, even though under existing federal law, passenger trains are supposed to have priority over freight trains. Other issues for Amtrak include the condition of the privately owned rail lines that can result in a local speed restriction below the track's normal speed, train breakdowns, and other incidents. Measures of these types of congestion problems are train on-time performance, amount of delay, and average speed.

In addition, as a type of passenger service, congestion problems with Amtrak theoretically may be manifest in ticket availability, ticket prices, and train overcrowding. Systemwide, these are generally not issues that Amtrak has to worry about. These problems may occur on certain routes at certain times, such as the NEC around major holidays, but realistically, the system cannot be designed to handle demand that only occurs a few times a year. Load factor, a metric tracked by Amtrak, is a measure of train utilization and possible overcrowding.

Intercity Passenger Rail (Amtrak) Congestion Trends

The data appear to show that, in general, rail system congestion, including freight, commuter, and Amtrak operations, is something of a problem and is getting worse, but that train overcrowding is not a problem. Amtrak delays per 10,000 miles have trended upward from FY2001 through FY2006. Delays resulting from Amtrak itself have remained relatively constant during that period, at about 400 minutes per 10,000 train miles. Most of the delays are due to freight operations, rising from about 1,700 minutes in FY2001 to about 2,300 minutes in FY2006. Overall on-time performance was 67.8% in FY2006, down from 69.8% in FY2005, 70.7% in FY2004, and 74.1% in FY2003. Load factors, on the other hand, are quite low, suggesting little train overcrowding. For all Amtrak routes, the load factor in FY2006 was 48%. The average load factor in FY2006 was 45% in the NEC, 41% in state-supported and other corridors, and 55% on long distance routes.[150]

THE COSTS OF TRANSPORTATION CONGESTION

The negative effects of transportation congestion are primarily economic. Transportation congestion, particularly road traffic congestion, also causes a good deal of stress in some of those that experience it, as well as a certain amount of environmental damage because of the extra fuel that is used. Congestion may also have a negative effect on road traffic safety, although it is not clear from the available evidence if the damage done as a result of slowing or stopped vehicles outweighs the reduction in crash severity due to lower speeds. However, the main effects are an increase in direct user costs, particularly the extra time and fuel expended, and a number of economic distortions that decrease productivity and hurt competitiveness.

Most of the available evidence on the costs associated with transportation congestion is limited to the effects of road traffic congestion. Little is known about the national costs associated with rail, transit, and intermodal congestion. Hence, if accurate, existing estimates focusing exclusively on the costs of road traffic congestion understate the total cost of transportation congestion to the national economy. It must also be borne in mind that estimates of the cost of congestion are based on assumptions that are somewhat arbitrary. Time, an important variable in transportation evaluation studies, can be especially hard to value.[151]

The direct user costs of road traffic congestion are the extra time and fuel expended to complete a trip. In its study of 437 cities, TTI estimates that drivers lost 4.2 billion hours to road traffic congestion and wasted an extra 2.9 billion gallons of fuel, at a cost of $78.2 billion.[152] Most of the cost is due to the time lost by travelers. Per traveler, the cost is $710 annually or approximately $3 per work day. In inflation-adjusted terms, the cost of congestion has risen from $14.9 billion in 1982 (in constant 2005 dollars). These estimates, however, do not include the cost of unreliability, in that travelers will often budget extra time to make sure they arrive on time, even if it means arriving early.

In addition to direct user costs, there are at least three other types of economic costs associated with congestion[153]:

- Logistics costs — the extra costs associated with businesses having to carry extra inventory as a result of slower and more unreliable transportation.
- Market scale and accessibility costs — as congestion reduces the area that can be served by a production facility, the reduced demand results in higher unit costs because of lower-scale efficiencies and lower access to specialized inputs.
- Business cost of worker commuting — the costs associated with attracting and retaining workers and compensating them for higher commuting costs. There may also be lower labor productivity resulting from the stress of longer or more unreliable commutes.

Although not quantified, congestion in other modes also has costs. As demand for space on the rail system increases, rates may begin to rise, increasing shipper costs. In addition, railroads have been keen to accommodate generally more lucrative intermodal shipments over bulk shipments. This is beginning to create significant problems for the movement of bulk shippers in some markets at certain times, as they often have no alternative to moving their goods by rail. Congestion on the rail system may also force more freight to move by truck. Some contend that there are a number of public benefits associated with moving freight by rail, such as less air pollution per ton-mile of freight than trucking.[154] Similarly, congestion and overcrowding in passenger rail transportation and public transportation may divert travelers to other modes. In urban areas, congested transit service may lead to more single-occupant driving during the peak period, causing more road congestion. Likewise, congested intercity rail transportation might shift a few travelers onto the roads, although it may shift them to intercity buses or airplanes, depending on the situation.

It is commonplace these days to attempt to quantify the costs of congestion and add them together to arrive at a total cost of congestion to the economy, sometimes expressed as a share of GDP. This approach is particularly common in accounting for the costs of road traffic congestion, as

TTI does in terms of extra time and fuel, and other researchers have attempted to calculate more comprehensively.[155]

There are, however, some problems with this approach. These cost estimates are often based on the premise of "free-flowing traffic," which, as discussed above, tends to exaggerate the amount of congestion experienced. Furthermore, total cost estimates suggest that there is a monetary windfall waiting to be distributed to every household, when in reality, eliminating congestion, if it were possible, would only save most travelers a few minutes on peak-period trips.[156] Consequently, a number of experts question the calculation of total costs and suggest that

> what matters in practical terms is the change in the cost of congestion brought about by a specific feasible projects or act of policy.... As economists would say, we need to change our thinking from total costs to marginal costs.[157]

TRANSPORTATION CONGESTION REMEDIES

Transportation engineers and planners have devised a large number of potential remedies for congestion. Although it is beyond the scope of this report to evaluate all of these, it is worthwhile discussing some of the major remedies as a basic guide for policy makers. The many different remedies form three basic strategies for reducing congestion: adding new capacity, operating the existing capacity more efficiently, and managing demand. This section discusses these strategies and the institutional issues that affect the implementation of congestion remedies. This is followed by a discussion of rail congestion remedies and intermodalism in freight transportation.

BUILDING NEW ROAD AND TRANSIT CAPACITY

Building new roads, or expanding existing ones, is one approach to reducing congestion. Proponents of road building point out that since the completion of the interstate system, road construction has generally lagged behind the growth in motor vehicle travel. Moreover, these proponents argue that in some places, lack of capacity is a major contributor to road congestion. TTI's analysis of congestion found that adding to road capacity slowed the growth in travel delay.[158] New capacity can range from major new freeways to major bottleneck reduction projects and much smaller projects, such as widening arterial roads and improving street connectivity.

Few deny that highway travel has grown more than highway capacity during the past few decades. There is, however, a major disagreement about whether new road capacity, in the absence of tolling pricing, can solve congestion because of the problem of induced demand (see earlier discussion). Other concerns about major expansions of road capacity have to do with the costs in labor and raw materials, rights-of-way acquisition in heavily developed urban areas, and social and environmental disruptions. Over the past few years, the cost of raw materials has increased dramatically, making this a greater concern than just a few years ago. An added difficulty is the time it takes to plan, design, and build major new facilities. Consequently, some experts argue that once congestion has developed, it is very hard for an area to build its way out of the problem because of the time it takes to add new capacity.

Some suggest that road congestion is a problem because other viable means of transportation are not widely available. In this view, new or expanded public transportation service is seen as a major solution to urban road traffic congestion. TTI points out that if public transit service disappeared and everyone used private vehicles, delay in the 437 urban areas it studied would increase by 541 million hours, about a 13% increase.[159] By its estimates, almost all of this extra delay (about 80%) would occur in very large urban areas (population of 3 million or more). This is because, as noted above, transit service is heavily concentrated in just a few major metropolitan areas. Currently, about 5% of workers commute by transit and in only the New York and Chicago metropolitan areas do more than 10% of commuters use transit. Nevertheless, much higher proportions of transit users are found for certain types of commute, particularly those from suburb to central city. It is probably in these sorts of situations — where the density of origins and destinations is high enough to make transit an attractive mode of travel — in which new or expanded transit options are likely to contribute to a reduction in road traffic congestion. Morever, because buses can be caught up in road traffic congestion, only dedicated bus lanes or non-highway modes of transit provide effective solutions. Generally speaking, transit is not likely to reduce congestion in smaller urban areas or in the suburbs of large urban areas because the areas to be covered are too large and the densities of residences and jobs too low.

According to some experts, new or expanded transit systems have improved travel options but have not noticeably reduced road traffic congestion.[160] To some extent, this is because most new major transit systems are built in fast-growing regions in which the growth in travel demand tends to swamp the extra capacity. However, some contend that peak-period road traffic congestion is not reduced because if some people switch from road to rail others are induced to travel by car at the most convenient times, or because many rail riders are not former drivers but former bus riders. Moreover, even though, theoretically, with more transit service, a greater number of people are able to travel at the most convenient times, the new capacity may not serve the greatest needs, such as suburb-to-suburb commutes.

Like new highway capacity, new transit capacity is costly in terms of labor, materials, and, in some cases, right-of-way acquisition. However, transit can have positive social and environmental benefits, such as potentially greater mobility for the poor and non-drivers, as well as lower air pollutant emissions per trip. New rail systems are the most costly, although light rail can be a cheaper alternative than heavy rail. The cost of new commuter rail capacity depends largely on whether or not the existing freight rail network is available for use by passenger trains. Because of the large start-up costs, some proponents of expanded transit capacity argue that new forms of bus transit, such as bus rapid transit (BRT), are a more viable alternative.

OPERATING EXISTING CAPACITY MORE EFFECTIVELY

Operational improvements on highways and transit have become a much more important concern of state and local DOTs as congestion has increased. Operations include a host of strategies for improving the flow of road traffic and improving transit trips. These include, among others, transportation management center operations, incident management techniques, event management techniques, ramp metering, real-time traveler information, road weather information systems, work zone management, signal retiming, and transit priority at signals. Many of these strategies rely on the deployment of Intelligent Transportation Systems (ITS) technologies.

In general, operational strategies for reducing congestion can be quicker to implement and relatively low-cost. For instance, with a large share of road traffic congestion caused by incidents and other non-recurring forms of delay, many areas have created transportation management centers to improve the response of state and local agencies to problems that can arise at any time or place in the transportation system. Evaluations have shown that in many cases, the benefits of these centers greatly outweigh the costs.[161] Another advantage of these types of programs is that they typically cause minimal disruptions, unlike major construction projects. On the downside, operational strategies require a much greater ongoing commitment from local and state DOTs. This has been a problem in some places because, historically, DOTs have functioned as road construction and maintenance agencies and have struggled to redefine their mission.

Managing Demand

Operational strategies reduce congestion on the supply side of the transportation equation. There are a range of strategies that exist on the demand side, known as demand management strategies. Among others, these include congestion (or value) pricing, high-occupancy vehicle (HOV) lanes, alternative work schedule and telecommuting programs, and land-use strategies. Proponents of demand management strategies argue that just as adding a few extra cars on a roadway can make a big difference in terms of extra delay, removing a few cars can make a big difference in terms of reducing delay. For example, an evaluation of the congestion charge in London, described below, suggests that while traffic has been reduced by about 15%, congestion has been reduced by about 30%.[162]

Congestion Pricing

Schemes to charge drivers a fee to travel on congested facilities or in congested areas are known generally as "congestion" or "value" pricing. Economists generally believe that congestion pricing is the single most viable way, though not necessarily most popular way, to reduce highway congestion. With the use of advanced technologies, the fee can be varied to ensure the most efficient use of the facility.

There are four main forms of road congestion pricing: variably priced lanes, variable tolls on entire roadways, cordon charges, and variable areawide charge pricing.[163] Cordon pricing, like the one instituted in London in 2003, charges a fee for entering an area at certain times. Facility-based pricing charges a fee to use a specific facility — usually a freeway or freeway lane — depending on the time of day and the amount of traffic on the facility. Variable areawide pricing would use some sort of vehicle tracking technology to charge for the amount of travel and the types of facilities used over an entire area.

The main advantage of congestion pricing is that demand can be managed to offer travel that is less likely to be subject to delay, especially unpredictable delay. Another advantage is that on existing roadways, congestion pricing can be implemented relatively quickly. Moreover, with congestion pricing, the negative external effects are minimal and the effects may even be positive, such as a reduction in air pollutant emissions from idling vehicles. For state and local governments, congestion pricing provides a revenue stream to pay for building and operating transportation facilities.

Congestion pricing schemes are often unpopular and have been criticized in a number of ways. One criticism is that they discriminate against low-income drivers. Although it is true that the toll will represent a greater burden for drivers with lower incomes, research has shown that low-income drivers do use tolled facilities, suggesting that they often value the time saved. Others propose that pricing facilities ought to be reserved for new capacity, particularly when it is made available alongside a typically congested but free alternative. Another criticism of congestion pricing is that by making it more expensive to travel downtown, the types of areas or facilities most likely to be tolled, businesses and consumers are likely to seek out locations away from the tolled areas, resulting in more sprawl. Some contend that, depending on how it is implemented, traffic may be diverted from the newly tolled facility to other roads that may be less well-equipped to deal with heavy volumes. Finally, some have argued that charging new tolls on an existing roadway is a form of double taxation because users have already financed the construction of the road through the gas tax and other user fees.

Land Use Strategies

It is often asserted that low density, suburban growth in housing and employment has contributed to road traffic congestion. Hence, some have suggested that one approach to congestion is to encourage different types of land use development that will reduce reliance on single-occupant vehicle travel. The two main types of land use strategies that are commonly proposed are (1) to encourage increased housing and employment density and (2) to improve the jobs/housing balance. The first often comes under the rubric of transit-oriented development, whereby more density will make transit, walking, and cycling more attractive transportation options. The second type of strategy does not necessarily entail alternatives to driving, but driving can be reduced when people live closer to where they work.

Although these are desirable strategies in many ways, experts point out one of the main disadvantages of them is that land-use patterns take decades to evolve, hence decisions taken today will take years to make a difference in the overall transportation/land-use system. Experience with increasing land-use densities shows that such strategies are not likely to reduce congestion per se, although they are likely to increase accessibility. In addition, some have suggested that such policies may raise the costs of developing housing, offices, and other types of facilities primarily by making land more expensive. Another disadvantage is that land-use decision making tends to be highly fragmented, so that policies to slow growth in one jurisdiction may lead to "leapfrog" development in another jurisdiction, causing more travel and more congestion. Research on improving the jobs/housing balance shows that it is unlikely to reduce congestion because, for a number of reasons, it is very difficult to get people to live near where they work.[164]

INSTITUTIONAL ISSUES

The problem of transportation congestion is compounded by the highly fragmented planning and operation of the transportation system. Most urban areas comprise numerous local governments, and some span multiple states. Important interstate corridors, like I-95, by definition suffer jurisdictional fragmentation. Even in a single jurisdiction, multiple agencies are responsible for different aspects of the system. Transit systems, for example, are often operated independently of local and state departments of transportation. Highway incidents may involve a whole host of agencies, including state police, local police, ambulance, fire, and state and local DOTs. In some cases, fragmentation involves a public-versus-private dimension. The rail system is mostly privately owed and operated, which can make it challenging to institute new passenger rail service, for example. Because of this fragmentation many anti-congestion strategies require coordination and collaboration functionally, jurisdictionally, and across the public/private divide.[165]

Efforts to promote voluntary coordination and collaboration between agencies and jurisdictions are typically uncontroversial. More controversial are solutions that affect the funding and authority of different jurisdictions in the planning and programming of transportation improvements. For instance, some have suggested that because congestion tends to occur on the regional scale, regional authorities, such as the Georgia Regional Transportation Authority, and metropolitan planning organizations should be given more power over the planning and operation of the transportation system vis-a-vis states and localities.[166]

EXPANDING RAIL CAPACITY

Building new capacity in freight rail is seen as a way of dealing with congestion issues, particularly as a host of technological changes that have improved operational throughput appear to have run their course.[167] In reasonable financial health today, freight railroads are investing to increase capacity.[168] However, there are concerns that this investment is not keeping up with demand.[169] A number of reasons have been proposed for this, many having to do with the uniqueness of freight railroading as an industry. To begin with, many note that because track and the accompanying operational systems are so costly, freight railroading is one of the most capital-intensive industries in America.[170] Also, once constructed, railroad track is fixed in space, representing a huge wager on future patterns of freight movement. It has been argued that similar risks are borne by the public sector in the trucking, air, and waterborne freight industries. Furthermore, like most infrastructure improvements, it takes a relatively long time to respond to market signals that may change quickly.

Another issue is whether or not railroads can be a solution to road traffic congestion by taking truck traffic off the highways. Clearly, rail will not be a solution to roadway congestion if there is insufficient rail capacity. But should public involvement for building rail capacity be predicated on relieving road congestion? As it stands today, there are a host of significant barriers to rail relieving road congestion, including the fact that many industrial facilities are no longer served by rail spurs, either because they have been built away from them or because the spurs were taken out during the downsizing of rail capacity.[171] However, the main reason that rail is unlikely to reduce urban road congestion is that in most places, trucks make up a small part of the traffic stream.[172] Nationally, trucks account for 8% of highway VMT,[173] although the effect of a truck on the traffic stream is greater than a passenger car. On a multilane highway with no grade, a large truck represents 1.7 cars and at intersections between 3 and 4 cars.[174]

Nevertheless, many support the idea of public funding for expanding rail capacity because it will improve the speed and efficiency of the freight system by allowing shipments to bypass urban road congestion. Moreover, many point out there are a range of public benefits to moving freight by rail, including less wear and tear on the roads and a possible reduction in air pollutant emissions. A corollary is that improved rail system capacity may also reduce the conflicts between freight and passenger trains, improving the speed and efficiency of both systems.

INTERMODALISM IN FREIGHT TRANSPORTATION

Many of the solutions for intermodal problems in freight transportation revolve around the connections to truck and rail transportation at water ports. The issues in these areas are particularly thorny because most ports are located in already congested urban areas with very limited space for expansion and because, as very large facilities in the freight system, they have a major impact on their physical and social environment in terms of pollution and noise, etc. Moreover, ports involve a complex mix of public and private organizations, blurring lines of responsibility and public and private benefits.

In this context, a number of improvements have been proposed to increase the speed with which freight moves through the system at these critical nodes without unduly affecting nearby residents. These improvements include extended truck gate hours, congestion pricing of dock facilities, truck appointment systems, expanded "on-dock" rail connections, truck-only lanes, and the development of inland ports connected by fast rail shuttles.

Concluding Observations

Like a number of other public policy issues, transportation congestion can be viewed as a collection of interrelated problems with severe constraints set in a context of continual change. These types of issues, sometimes called "wicked problems" in some, mostly non-transportation, public policy circles, often seem intractable and typically engender a good deal of frustration that nothing is being done or that what is being done is ineffectual at best or counterproductive at worst.[175] Among other things, these types of problems typically have other characteristics such as no definitive definition; a wide variety of potential solutions, but intense disagreements about the preferred ones and about what constitutes success; and, because of intended and unintended consequences, a situation where each solution tends to modify the problem in such a way as to make it manifest in a different form or in a different time or place.

When tackling these types of problems, public policy experts advise that the traditional linear approach, in which data are gathered, analyzed, and a solution formulated and implemented, is not workable. By contrast, they suggest that "solving" wicked problems is usually an ongoing, complex, and chaotic struggle that often requires incorporating multiple viewpoints and approaches at once. Moreover, these experts note that when working on solutions, policy makers and planners often encounter new dimensions of the problem and, therefore, they must be creative and opportunity-driven.[176]

In terms of transportation congestion, this suggests a few key ideas for policy makers to keep in mind as solutions to this problem are crafted and pursued in the future. To begin with, there is no one solution that will ever fully solve transportation congestion and that, paradoxically, fully solving the problem may be undesirable because congestion can be a good problem to have in some circumstances and may also be a choice about how to distribute scarce resources. Another key idea is that each solution applied to a dimension of transportation congestion might create other unintended problems along other dimensions that require new creative solutions. As such, multiple, iterative strategies likely will be needed, including supply-side and demand-side approaches; approaches that focus on passenger systems and those that focus on freight, highway strategies, and transit strategies; and those that promise short-term results and some that promise improvement in the long-term. Possibly and most importantly, policy makers and planners should consider that the ultimate goal may not be to reduce or eliminate transportation congestion per se, but to focus instead on improving passenger and freight mobility and accessibility.

REFERENCES

[1] U.S. Department of Transportation, "National Strategy to Reduce Congestion on America's Transportation Network," May 2006, at [http://www.fightgridlocknow.gov/docs/conginitoverview070201.htm].

[2] Transportation Research Board, *Critical Issues in Transportation* (Washington, DC, 2006), p. 2, at [http://onlinepubs.trb.org/onlinepubs/general/CriticalIssues06.pdf].

[3] U.S. Government Accountability Office, *High Risk Series: An Update*, GAO-07-310, January 2007, at [http://www.gao.gov/new.items/d05207.pdf].

[4] See U.S. Department of Transportation, Federal Highway Administration, *America's Highways, 1776-1976* (Washington, DC, 1976), especially Part Two, Chapter 1.

[5] *Transportation Weekly*, "Congress Completes Work on Highway Bill," vol. 6, issue 34, August 4, 2005.

[6] Texas Transportation Institute, *Urban Mobility Report 2007* (College Station, TX, 2007), at [http://mobility.tamu.edu/ums/].

[7] U.S. Congressional Budget Office, *Freight Rail Transportation: Long-Term Issues*, January 2006, at [http://www.cbo.gov/showdoc.cfm?index=7021&sequence=0].

[8] Downs, Anthony, *Still Stuck in Traffic: Coping with Peak-Hour Traffic Congestion* (Washington, DC: Brookings Institution Press, 2006).

[9] Taylor, Brian D., "Rethinking Congestion," *Access*, vol. 21 (2002), pp. 8-16.

[10] Ibid.; Downs, 2006; El-Geneidy, Ahmed M. and David M. Levinson, "Access to Destinations: Development of Accessibility Measures,"

report prepared for the Minnesota Department of Transportation, May 2006, at [http://www.lrrb.org/pdf/200616.pdf].

[11] See, for instance, Sierra Club, *Highway Health Hazards* (San Francisco, CA, 2004), at [http://www.sierraclub.org/sprawl/report04_highwayhealth/report.pdf].

[12] CRS calculation based on U.S. Bureau of Economic Analysis, "Personal Income for Metropolitan Areas, 2006," Table 1, News Release, August 7, 2007, at [http://www.bea.gov].

[13] Crafts, Nicholas and Timothy Leunig, "The Historical Significance of Transport for Economic Growth and Productivity," background paper for the Eddington Transport Study, October 2005, at [http://www.hm-treasury.gov.uk/independent_reviews/eddington_transport _study/eddington_index.cfm].

[14] See the maps in U.S. Department of Transportation, Federal Highway Administration, *Freight Facts and Figures 2007* (Washington, DC, 2007), pp. 31-32, at [http://ops.fhwa.dot.gov/freight/freight_analysis/nat_freight_stats/docs/07factsfigures/in dex.htm].

[15] U.S. Census Bureau, *Statistical Abstract of the United States, 2008* (Washington, DC, 2007), p. 8, at [http://www.census.gov/compendia/statab/].

[16] U.S. Congressional Budget Office, *The Long Term Budget Outlook: Supplemental Datasheet* (Washington, DC, December 2007), at [http://www.cbo.gov/ftpdocs/88xx/doc8877/SupplementalData.xls].

[17] Federal Highway Administration, 2007, p. 11.

[18] See CRS Report RL31735, *Federal-Aid Highway Program: "Donor-Donee" State Issues*, by Robert S. Kirk.

[19] For a discussion of these three elements in the early 1990s, see CRS Report 93-107, *Transportation Infrastructure: Economic Issues and Public Policy Alternatives*, by J.F. Hornbeck. (Out of print; available from the author.)

[20] See, for instance, American Society of Civil Engineers, "Report Card for America's Infrastructure 2005," at [http://www.asce.org/reportcard/2005/page.cfm?id=30].

[21] U.S. Department of Transportation, Federal Highway Administration and Federal Transit Administration, *2006 Status of the Nation's Highways, Bridges, and Transit: Conditions and Performance* (Washington, DC, 2007), at [http://www.fhwa.dot.gov/policy/2006cpr/index.htm].

[22] Ibid., exhibit 6-11.

References

[23] Ibid., exhibit 6-8. The federal share of highway spending as a whole is lower, currently about 22% of spending by all levels of government. The same pattern of a shrinking federal share since 1980 is similar, however.

[24] Ibid., exhibit 6-22; American Public Transportation Association, "Unlinked Passenger Trips by Mode, 1890-2004," at [http://www.apta.com/research/stats/ridership/trips.cfm].

[25] Ibid., exhibit 6-20.

[26] Ibid., exhibit 6-23.

[27] Ibid., exhibits 6-8, 6-23.

[28] Ibid., exhibits 6-8, 6-20.

[29] Ibid., exhibit 3-4.

[30] Ibid., exhibit 3-18.

[31] Ibid., exhibit 4-12.

[32] Ibid., exhibits 3-24, 3-28, 4-15, 4-18.

[33] Testimony of Carl D. Martland, Senior Research Associate, Massachusetts Institute of Technology, in U.S. Congress, House Committee on Transportation and Infrastructure, Subcommittee on Railroads, U.S. Rail Capacity Crunch, April 26, 2006.

[34] American Association of State Highway and Transportation Officials (AASHTO), *Transportation, Invest in America: Freight-Rail Bottom Line Report* (Washington, DC, 2003), at [http://freight.transportation.org/doc/FreightRailReport.pdf].

[35] Puentes, Robert and Linda Bailey, "Increasing Funding and Accountability for Metropolitan Transportation Decisions," in Bruce Katz and Robert Puentes, eds., *Taking the High Road: A Metropolitan Agenda for Transportation Reform* (Washington, DC: Brookings Institution Press, 2005).

[36] Edward Hill et al., "Slanted Pavement: How Ohio's Highway Spending Shortchanges Cities and Suburbs," in Katz and Puentes, 2005.

[37] Downs, Anthony and Robert Puentes, "The Need for Regional Anticongestion Policies," in Katz and Puentes, 2005; Lewis, Paul G., "Regionalism and Representation: Measuring and Assessing Representation in Metropolitan Planning Organization," *Urban Affairs Review*, vol. 33, no. 6 (July 1998), pp. 839-853; Association of Metropolitan Planning Organizations, "AMPO Survey Results: Policy Board Structure," at
[http://www.ampo.org/assets/62_policyboardstructure.doc].

[38] U.S. General Accounting Office, *Surface Transportation: Many Factors Affect Investment Decisions*, GAO-04-744 (Washington, DC, June 2004), at [http://www.gao.gov/new.items/ d04744.pdf].

[39] U.S. General Accountability Office, *Highway and Transit Investments: Options for Improving Information on Projects' Benefits and Costs for Increasing Accountability for Results*, GAO-05-172 (Washington, DC, January 2005), at [http://www.gao.gov/new.items/ d05172.pdf].

[40] HM Treasury and Department for Transport, *The Eddington Transport Study, Executive Summary* (London, December 2006), p. 6, at [http://www.hm-treasury.gov.uk/media/39A/41/eddington_execsum 11206.pdf].

[41] Cox, Wendell, Alan E. Pisarski, and Ronald D. Utt, "Rush Hour: How States Can Reduce Congestion Through Performance-Based Transportation Programs," *Heritage Foundation Backgrounder*, no. 1995 (January 10, 2007), at [http://www.heritage.org/Research/Smart Growth/upload/bg_1995.pdf].

[42] U.S. General Accounting Office, June 2004.

[43] Surface Transportation Policy Project, *Easing the Burden: A Companion Analysis of the Texas Transportation Institute's 2001 Urban Mobility Study* (Washington, DC, May 2001), at [http://www.transact.org/PDFs/etb_report.pdf].

[44] Wendell Cox and Randal O'Toole, "The Contribution of Highways and Transit to Congestion Relief: A Realistic View," *Heritage Foundation Backgrounder*, no. 1721 (January 27, 2004).

[45] Pickerell, Don, "Induced Demand: Definition, Measurement and Significance," in *Working Together to Address Induced Demand* (Washington, DC: Eno Transportation Foundation, 2002).

[46] For a discussion of this issue see Downs, 2006, p. 104.

[47] Pickerell, 2002.

[48] Cervero, Robert, "Are Induced-Travel Studies Inducing Bad Investments?," *Access*, no 22, 2003, pp. 22-27, at [http://www.uctc.net/access/22/Access%2022%20-% 2004%20-%20Induced%20Travel%20Studies.pdf].

[49] Downs, 2006; Poole, Robert, "New Evidence Questions the Reality of 'Induced Demand,'" *Surface Transportation Innovations*, no. 39 (January 2007), at [http://www.reason.org/surfacetransportation39.shtml].

[50] Transportation Research Board, *Fare Policies, Structures, and Technologies: Update*, Transit Cooperative Research Program

(TCHRP) Report 94 (Washington, DC, 2003), table 2-6, at [http://onlinepubs.trb.org/onlinepubs/tcrp/tcrp_rpt_94.pdf].
[51] CBO, 2006.
[52] U.S. Government Accountability Office (GAO), *Freight Railroads: Industry Health Has Improved, but Concerns about Competition and Capacity Should Be Addressed*, GAO-07-94 (Washington, DC, October 2006), at [http://www.gao.gov/new.items/d0794.pdf]; AASHTO, 2003.
[53] Bryan, Joseph, Glen Weisbrod, and Carl Martland, *Assessing Rail Freight Solutions to Roadway Congestion: Final Report*, NCHRP Project 8-42 (Washington, DC: Transportation Research Board, October 2006), at [http://www.trb.org/NotesDocs/NCHRP08-42_FR_Rev 10-06.pdf].
[54] CBO, 2006.
[55] Ibid., p.20.
[56] Ibid.
[57] Illinois Commerce Commission, "Motorist Delay at Public Highway-Rail Grade Crossings in Northeastern Illinois," Working Paper 2002-03 (July 2002), at [http://www.icc.illinois.gov/docs/rr/021114 rrdelay.pdf].
[58] See, for a general overview, U.S. Department of Transportation, Federal Highway Administration, "Evaluating the Performance of Environmental Streamlining: Development of a NEPA baseline for Measuring Continuous Performance" (Washington, DC), at [http://www.environment.fhwa.dot.gov/strmlng/baseline/index.asp].
[59] Wachs, M., "Fighting Traffic Congestion with Information Technology," *Issues in Science and Technology* (Fall 2002), at [http://issues.org/19.1/wachs.htm].
[60] Motor vehicle ownership increased from approximately 8,000 in 1900 to 27 million in 1930. See U.S. Department of Transportation, Federal Highway Administration, *Highway Statistics, Summary to 1985* (Washington, DC, 1987), p. 25.
[61] American Public Transportation Association, *2006 Public Transportation Fact Book* (Washington, DC, 2006), at [http://www.apta.com/research/stats/factbook/index.cfm].
[62] Weingroff, Richard F., "The Genie in the Bottle: The Interstate System and Urban Problems, 1939-1957," *Public Roads*, vol. 64, no. 2 (September/October 2000), pp. 2-15, at [http://www.tfhrc.gov/pubrds/septoct00/urban.htm].

[63] In a speech Eisenhower argued, "The country urgently needs a modernized interstate highway system to relieve existing congestion, to provide for the expected growth of motor vehicle traffic, to strengthen the Nation's defenses, to reduce the toll of human life exacted each year in highway accidents, and to promote economic development." Quoted in Weingroff, R.D., "Original Intent: Purpose of the Interstate System: 1954-56," at [http://www.fhwa.dot.gov/infrastructure/originalintent.cfm], as of December 28, 2006.

[64] Schwartz, Gary T., "Urban Freeways and the Interstate System," *Southern California Law Review*, vol. 49 (1976), pp. 406-513.

[65] U.S. Department of Transportation, 1976, p. 481.

[66] Paved centerline miles is the length of roads paved with some type of bituminous, Portland cement concrete, or brick surface, as measured along the center in one direction. As such, this metric does not account for the capacity provided by highways with more than one lane in each direction. FHWA did not begin publishing lane-mile data until 1980.

[67] Smerk, George M., *The Federal Role In Urban Mass Transportation* (Bloomington, IN: Indiana University Press, 1991).

[68] American Public Transportation Association, "Transit Ridership Report, Fourth Quarter 2005," April 4, 2006, at [http://www.apta.com/research/stats/ridership/riderep/documents/05q4cvr.pdf].

[69] Polzin, Steven, and Xuehao Chu, *Public Transit in America: Results from the 2001 National Household Travel Survey* (Washington, DC, September 2005), at [http://www.nctr.usf.edu/pdf/527-09.pdf].

[70] Sun, Lena H., "Rush Hour Metro Fares May Rise As Much as $2.10," *The Washington Post*, December 14, 2006, p. A1.

[71] The Surface Transportation Board, the federal agency responsible for economic regulation of the railroad industry, classifies freight railroads based on operating revenue. In 2006, the classification was as follows: Class I, $346.8 million or more; Class II, $27.8 million to $346.7 million; Class III, less than $27.7 million.

[72] Association of American Railroads, *Railroad Facts 2007* (Washington, DC), p. 45.

[73] Ibid., p. 27.

[74] U.S. Department of Transportation, Federal Highway Administration, "Freight Carriers: From Modal Fragmentation to Coordinated Logistics," undated white paper, p. 1, at [http://ops.fhwa.dot.gov/freight/theme_papers/final_thm5_v4.htm].

[75] U.S. Bureau of Economic Analysis, National Income and Product Accounts, at [http://www.bea.gov].

[76] U.S. Army Corps of Engineers, *Waterborne Commerce Statistics of the United States 2005, National Summaries* (New Orleans, LA, 2007), p. 1-3, at [http://www.iwr.usace.army.mil/ndc/wcsc/pdf/wcusnatl05.pdf]; U.S. Army Corps of Engineers, *2006 Preliminary Waterborne Commerce Statistics: National Totals and Selected Inland Waterways* (New Orleans, LA, October 23, 2007), p. 1, at [http://www.iwr.usace.army.mil/ndc/wcsc/pdf/Prelim06.pdf].

[77] U.S. Department of Transportation, Federal Highway Administration, "Regulation: From Economic Deregulation to Safety Regulation," undated white paper, at [http://ops.fhwa.dot.gov/freight/theme_papers/final_thm8_v4.htm].

[78] Smerk, 1991.

[79] U.S. Congress, Senate Committee on Environment and Public Works, S.Rept. 102-71, June 4, 1991; U.S. Congress, House Committee on Public Works and Transportation, H.Rept. 102-171(I), July 26, 1991.

[80] S.Rept. 102-71, June 4, 1991, p. 4.

[81] At this time, the Interstate Highway System was approximately 45,000 miles in length. The interstates were part of the Federal-Aid Primary System, which also included approximately 260,000 miles of mostly rural arterials and some urban principal arterials. The Strategic Highway Network (STRAHNET) was, and still is, a system of roadways identified as being important for national defense. In addition to the interstates, the STRAHNET at this time included another 15,000 miles of non-interstate roads.

[82] U.S. Congress, H.Rept. 102-171(I), July 26, 1991, p. 6.

[83] Ibid., p. 7.

[84] Transportation Research Board, *The Congestion Mitigation and Air Quality Improvement Program: Assessing 10 Years of Experience*, Special Report 264 (Washington, DC, 2002).

[85] U.S. Government Accounting Office, *Transportation Infrastructure: States' Implementation of Transportation Management Systems*, GAO-RCED-97-32 (Washington, DC, 1997), at [http://www.gao.gov/archive/1997/rc97032.pdf].

[86] U.S. Department of Transportation, *Transportation Equity Act for the 21st Century — A Summary* (Washington, DC, 1998).

[87] U.S. Department of Transportation, Federal Motor Carrier Safety Administration, *Introductory Guide to CVISN* (Washington, DC,

2000), at [http://cvisn.fmcsa.dot.gov/ downdocs/cvisndocs/guides/intro_p2/pdf_all1/intro_p2full.pdf].

[88] ISTEA had identified 21 high priority corridors, and the NHS Designation Act had added another 8 corridors. ISTEA provided funds for feasibility and design studies.

[89] See CRS Report RL32226, *Highway and Transit Program Reauthorization Legislation in the 2nd Session of the 108th Congress*, by John W. Fisher.

[90] U.S. Congress, House Committee on Transportation and Infrastructure, H.Rept. 108-452, March 29, 2004.

[91] A similar provision was included in S. 1072.

[92] CRS Report RL33119, *Safe, Accountable, Flexible, Efficient Transportation Equity Act — A Legacy for Users: Selected Major Provisions*, coordinated by John W. Fisher.

[93] Ibid.

[94] Giglio, Joseph M., *Mobility: America's Transportation Mess and How to Fix It* (Washington, DC: Hudson Institute, 2005).

[95] Taylor, 2002.

[96] El-Geneidy, Ahmed M. and David M. Levinson, May 2006.

[97] Transportation Research Board, *Quantifying Congestion, Volume 1*, National Cooperative Highway Research Program, Report 398 (Washington, DC, 1997).

[98] Texas Transportation Institute, *Urban Mobility Report 2007* (College Station, Texas), Appendix A, at [http://mobility.tamu.edu/ums/].

[99] Texas Transportation Institute and Cambridge Systematics, *Monitoring Urban Freeways in 2003*, report prepared for the U.S. Department of Transportation, Federal Highway Administration, December 2004, at [http://tti.tamu.edu/documents/FHWA-HOP-05-018.pdf].

[100] American Highway Users Alliance, *Unclogging America's Arteries: Effective Relief for Highway Bottlenecks, 1999-2004* (Washington, DC, February 2004), at [http://www.highways.org/pdfs/bottleneck2004.pdf].

[101] Cambridge Systematics, "An Initial Assessment of Freight Bottlenecks on Highways," report prepared for U.S. Department of Transportation, Federal Highway Administration, October 2005, at [http://www.fhwa.dot.gov/policy/otps/bottlenecks/bottlenecks.pdf].

[102] Texas Transportation Institute and Battelle Memorial Institute, "International Border Crossing Truck Travel Time for 2001," report prepared for U.S. Department of Transportation, Federal Highway

Administration, April 2002, at [http://ops.fhwa.dot.gov/freight/documents/brdr_synthesis.pdf].

[103] U.S. Department of Transportation, Federal Highway Administration, Office of Freight Management and Operation, *The Freight Story: A National Perspective on Enhancing Freight Transportation* (Washington, DC, November 2002), p. 13, at [http://ops.fhwa.dot.gov/freight/freight_analysis/freight_story/freight.pdf].

[104] U.S. Department of Transportation, Federal Highway Administration, Office of Freight Management and Operation, *Freight Performance Measurement: Travel Time in Freight Significant Corridors*, FHWA-HOP-07-071, December 2006, at [http://ops.fhwa.dot.gov/freight/freight_analysis/perform_meas/fpmtraveltime/traveltime brochure.pdf].

[105] Goodwin, Phil, "The Economic Costs of Road Traffic Congestion," Discussion Paper, Transport Studies Unit, University College London, 2004, p. 13, at [http://eprints.ucl.ac.uk/archive/00001259/01/2004_25.pdf].

[106] Ibid.

[107] Downs, 2006, Appendix A.

[108] Oak Ridge National Laboratory, *Temporary Losses of Highway Capacity and Impacts on Performance: Phase 2* (Oak Ridge, TN, October 2004), at [http://www-cta.ornl.gov/cta/Publications/tlc/tlc2_title.shtml].

[109] Cambridge Systematics and Texas Transportation Institute, "Traffic Congestion and Reliability: Trends and Advanced Strategies for Congestion Mitigation," report prepared for U.S. Department of Transportation, Federal Highway Administration (September 1, 2005), at [http://ops.fhwa.dot.gov/congestion_report/congestion_report_05.pdf].

[110] Texas Transportation Institute, 2007.

[111] U.S. Department of Transportation, Federal Highway Administration, *Highway Statistics* (Washington, DC, Annual Issues), at [http://www.fhwa.dot.gov/policy/ohpi/hss/index.htm].

[112] U.S. Census Bureau, *Statistical Abstract of the United States, 2008* (Washington, DC, 2007), pp. 7, 373; U.S. Department of Transportation, Research and Innovative Technology Administration, *National Transportation Statistics 2007* (Washington, DC, 2007), table 1-11; U.S. Bureau of Economic Analysis, "Gross Domestic Product," at [http://www.bea.gov/].

[113] Pisarski, Alan E., *Commuting in America III* (Washington, DC, Transportation Research Board, 2006). Of the 99.1 million commutes originating in a metropolitan area in 2000, 44.3 million (45%) were from suburb to suburb, 25.2 million central city to central city (25%), 18.8 million from suburb to central city (19%), 8.6 million from central city to suburb (9%), and 2.1 million to a non-metropolitan destination (2%).

[114] Pisarski, 2006; U.S. Census Bureau, *2005 American Community Survey*, at [http://www.census.gov/].

[115] U.S. Census Bureau and U.S. Department of Housing and Urban Development, *American Housing Survey for the United States: 2005* (Washington, DC, 2006), table 2-7, at [http://www.census.gov/prod/2006pubs/h150-05.pdf].

[116] American Petroleum Institute, "U.S. Pump Price Update — April 10, 2007," at [http://www.api.org/aboutoilgas/gasoline/upload/PumpPriceUpdate.pdf].

[117] FHWA, 2007, p. 20.

[118] U.S. Department of Transportation, Federal Highway Administration, *Highway Statistics 2006* (Washington, DC, 2007b).

[119] U.S. Census Bureau, 2007, p. 8.

[120] Congressional Budget Office, December 2007.

[121] Federal Highway Administration and Federal Transit Administration, 2007.

[122] Ibid., pp. 9-10. Rural VMT is projected to grow faster than urban VMT for several reasons: urban areas, unlike rural areas, are expected to moderate their VMT growth using travel demand management techniques; commercial truck travel in rural areas is expected to grow more quickly than in urban areas; and rural areas include rapidly growing places on the urban fringe that may be reclassified as urban in the future.

[123] Federal Highway Administration, 2007, p. 11.

[124] U.S. Department of Transportation. Federal Highway Administration, Office of Freight Management and Operation, 2006.

[125] CRS calculations based on U.S. Department of Transportation, Federal Highway Administration, *Highway Statistics* (Washington, DC, annual issues).

[126] See the maps in Federal Highway Administration, 2007, pp. 31-32.

[127] Cambridge Systematics and Texas Transportation Institute, "Traffic Congestion and Reliability: Linking Solutions to Problems," report prepared for U.S. Department of Transportation, Federal Highway

References

Administration (July 19, 2004), p. 2-1, at [http://ops.fhwa.dot.gov/congestion_report_04/congestion_report.pdf].
[128] American Highway Users Alliance, 2004.
[129] U.S. Department of Transportation, Federal Highway Administration, "An Initial Assessment of Freight Bottlenecks on Highways," white paper prepared by Cambridge Systematics, October 2005, at [http://www.fhwa.dot.gov/policy/otps/bottlenecks/index.htm].
[130] American Highway Users Alliance, American Automobile Association and TRIP, "Are We There Yet?" (Washington, DC, 2005), at [http://www.highways.org/pdfs/travel_ study2005.pdf].
[131] Federal Highway Administration, 2007, p. 14.
[132] Ibid., p. 11.
[133] U.S. Department of Transportation, Research and Innovative Technology Administration, Bureau of Transportation Statistics, *Transportation Statistics Annual Report 2005* (Washington, DC, 2005), at [http://www.bts.gov/publications/transportation_statistics_annual_report/2005/].
[134] Texas Transportation Institute and Battelle Memorial Institute, "International Border Crossing Truck Travel Time for 2001," report prepared for U.S. Department of Transportation, Federal Highway Administration (April 2002), at [http://ops.fhwa.dot.gov/freight/documents/brdr_synthesis.pdf].
[135] Taylor, John C., Douglas R. Robideaux, and George C. Jackson, "U.S.-Canada Transportation and Logistics: Border Impacts and Costs, Causes, and Possible Solution," *Transportation Journal*, vol. 43, no. 4, pp. 5-21.
[136] Testimony of Margaret Wrightson, Director of Homeland Security and Justice Issues, Government Accountability Office, in U.S. Congress, Senate Committee on Commerce, Science and Transportation, May 17, 2005, at [http://www.gao.gov/new.items/d05448t.pdf].
[137] Transportation Research Board, *Transit Capacity and Quality of Service Manual, 2nd Edition*, TCRP Report 100 (Washington, DC, 2003), at [http://nrc40.nas.edu/news/blurb_detail.asp?id=2326].
[138] American Public Transportation Association, "Unlinked Passenger Trips by Mode, 1890-2005," at [http://www.apta.com/research/stats/ridership/trips.cfm].
[139] Federal Highway Administration and Federal Transit Administration, 2007.
[140] Ibid., pp. 4-3

[141] Ibid., pp. 4-22.
[142] Ibid., exhibit 4-15.
[143] Association of American Railroads, *Railroad Facts 2007* (Washington, DC, November 2007), p. 42.
[144] Transportation Research Board, *Freight Capacity for the 21st Century*, Special Report 271 (Washington, DC, 2003), p. 62.
[145] Association of American Railroads, 2007.
[146] CBO, January 2006, p. 8.
[147] Testimony of Carl D. Martland, April 26, 2006.
[148] CRS calculations using the implicit price deflator for GDP.
[149] Association of American Railroads, *National Rail Freight Infrastructure Capacity and Investment Study*, Washington, DC, September 2007, at [http://www.aar.org/PubCommon/Documents/natl_freight_capacity_study.pdf]
[150] Amtrak, "Monthly Performance Report for September 2006," December 4, 2006, at [http://www.amtrak.com/pdf/0609monthly.pdf]; Amtrak, "Monthly Performance Report for September 2004," November 1, 2004, at [http://www.amtrak.com/pdf/0409monthly.pdf].
[151] U.S. Department of Transportation, "Departmental Guidance for the Valuation of Travel Time in Economic Analysis," memorandum, April 9, 1997; and U.S. Department of Transportation, "Revised Departmental Guidance," memorandum, February 11, 2003, at [http://ostpxweb.dot.gov/policy/programsa.htm#V].
[152] TTI assumes a cost of $14.60 per hour of person travel and $77.10 per hour of truck time. Excess fuel is estimated using the state average cost.
[153] Transportation Research Board, *Economic Implications of Congestion*, National Cooperative Highway Research Program, Report 463 (Washington, DC, 2001), at [http://onlinepubs.trb.org/onlinepubs/nchrp/nchrp_rpt_463-a.pdf].
[154] AASHTO, 2003.
[155] For example, the Chief Economist of the U.S. Department of Transportation adds TTI's estimate for 85 urban areas contained in the 2005 Urban Mobility Report with the cost of urban areas not included and other factors to arrive at a total of $168 billion annually. See Wells, Jack, "The Role of Transportation in the U.S. Economy," PowerPoint presentation to the National Surface Transportation Policy and Revenue Study Commission (June 26, 2006), slide 21, at [http://www.transportationfortomorrow.org/pdfs/

References

commission_meetings/0606_meeting_washington/wells_presentatio n_0606_meeting.pdf].
[156] Downs, 2006.
[157] Goodwin, 2004, p. 14.
[158] Texas Transportation Institute, 2007.
[159] Ibid.
[160] Downs, 2006.
[161] U.S. Department of Transportation, Joint Program Office, *Intelligent Transportation Systems Benefits, Costs and Lessons Learned: 2005 Update* (Washington, DC, May 2005), at [http://www.itsdocs.fhwa.dot.gov/JPODOCS/REPTS_TE/14073_files/14073.pdf].
[162] Transport for London, "Central London Congestion Charging: Impacts Monitoring, Fourth Annual Report," June 2006, at [http://www.tfl.gov.uk/assets/downloads/corporate/FourthAnnualReportFinal.pdf].
[163] U.S. Department of Transportation, Federal Highway Administration, "Congestion Pricing: A Primer," December 2006.
[164] Downs, 2006.
[165] U.S. Department of Transportation, Federal Highway Administration, *Regional Transportation Operations Collaboration and Coordination: A Primer for Working Together to Improve Transportation Safety, Reliability and Security*, FHWA-OP-03-008 (Washington, DC, 2003), at [http://www.itsdocs.fhwa.dot.gov//JPODOCS/REPTS_TE//13686/13686.pdf].
[166] Puentes and Bailey, 2005; Downs and Puentes, 2005.
[167] Positive Train Control, a way of managing the movement of trains with advanced information and communications technologies, which has yet to be applied on a large scale, may have a significant impact on throughput, as well as other operational factors such as safety.
[168] GAO, October 2006.
[169] Testimony of Federal Railroad Administrator Joseph H. Boardman, in U.S. Congress, House Subcommittee on Railroads, Committee on Transportation and Infrastructure, The U.S. Rail Capacity Crunch, April 26, 2006.
[170] CBO, 2006.
[171] Bryan et al., 2006, p. 114.
[172] Bryan et al., 2006.
[173] Federal Highway Administration, 2007.
[174] Bryan et al., 2006.

[175] Horst Rittel and Melvin Webber, "Dilemmas in a General Theory of Planning," *Policy Sciences*, vol. 4 (1973), pp. 155-169.

[176] Conklin, Jeffery E. and William Weil, "Wicked Problems: Naming the Pain in Organizations," Touchstone Consulting Group, undated white paper, at [http://www.touchstone.com/tr/wp/wicked.html].

INDEX

A

abatement, viii, 2
access, viii, 2, 3, 8, 9, 12, 21, 25, 26, 33, 34, 38, 48, 54, 70
accessibility, vii, 1, 3, 34, 35, 54, 62, 66
accidents, 48, 72
accounting, 38, 54
administration, 29
age, 41
aggregates, 36
aid, 3, 27, 28, 30
air pollution, vii, 1, 3, 8, 26, 54
air quality, 20, 27
airplanes, 54
airports, vii, 1, 2, 4, 8, 23, 43
alternative(s), viii, 2, 4, 8, 9, 10, 13, 14, 15, 17, 54, 59, 60, 61
ambulance, 62
American Public Transportation Association, 69, 71, 72, 77
analysts, 22
argument, 12, 15, 25
Arizona, 41, 51
Army Corps of Engineers, 73
assessment, 7, 10, 13, 27
assets, 17, 69, 79
Association of American Railroads, 48, 49, 50, 51, 72, 78
assumptions, 10, 53
attention, 25
attractiveness, 41
authority, 31, 47, 62
availability, 40, 41, 51

B

baby boomers, 42
barges, 16
barriers, 23, 63
benefits, 8, 12, 13, 14, 17, 29, 54, 59, 63
bonds, 28, 32
border crossing, 7, 23, 36, 41, 43
Boston, 21, 45
bottleneck(s), 7, 8, 12, 32, 34, 36, 42, 43, 48, 51, 57, 74, 77
breakdown, 45
bumper-to-bumper, 33
bundling, 23
bus lanes, 21, 58
bus rapid transit, 59
bypass, 63

Index

C

California, 8, 36, 41, 72
Canada, 43, 77
capacity, viii, 2, 3, 11, 13, 14, 16, 20, 22, 23, 25, 26, 27, 30, 32, 33, 34, 35, 37, 38, 40, 42, 44, 46, 47, 48, 49, 51, 57, 58, 59, 61, 63, 69, 71, 72, 75, 77, 78, 79
capacity building, 25
capital, 10, 11, 17, 26, 30, 63
capital expenditure, 11
capital markets, 11, 17
carrier, 34
cars, vii, 1, 3, 4, 8, 19, 22, 41, 43, 60, 63
cement, 20, 72
Census Bureau, 41, 45, 68, 75, 76
central city, 13, 58, 76
chaotic, 65
Chicago, 11, 21, 45, 51, 58
Chicago Skyway, 11
classification, 72
Clean Air Act Amendments (CAAA), 26, 27
collaboration, 62
College Station, 40, 67, 74
commerce, 7
commercial, 29, 38, 41, 76
commodities, 48
community, 17, 27
commuter, 7, 21, 23, 44, 45, 46, 47, 49, 52, 59
competition, 22, 47
competitiveness, 53
complexity, 41, 48
compliance, 17
concentration, vii, 1, 7
concrete, 20, 72
conflict, 23
congestion, 1, 2, 15, 19, 25, 27, 28, 29, 30, 31, 32, 33, 34, 35, 38, 39, 40, 42, 43, 44, 47, 48, 51, 52, 53, 54, 57, 60, 61, 67, 70, 71, 73, 74, 75, 76, 78, 79
congestion pricing, viii, 2, 4, 9, 15, 60, 61, 64
congestion-free, viii, 2, 8, 9, 37
Congress, vii, viii, 1, 2, 4, 7, 9, 12, 13, 20, 30, 31, 67, 69, 73, 74, 77, 79
Congressional Budget Office (CBO), 7, 9, 41, 49, 67, 68, 71, 76, 78, 79
connectivity, viii, 2, 3, 9, 25, 51, 57
consensus, 13
conservation, 3
consolidation, 23
constraints, 65
construction, 20, 26, 30, 45, 57, 59, 61
consumers, 8, 61
consumption, 43
control, 19, 23, 26
cost-effective, 13
costs, 4, 8, 12, 13, 14, 16, 17, 23, 29, 33, 37, 53, 54, 58, 59, 62
covering, 39
credentials, 29
credit, 29
criticism, 34, 36, 37, 61
CRS, 1, 31, 68, 74, 76, 78
current ratio, 17
customers, 8
cycling, 61

D

dating, 22
deaths, 3
decision making, 62
decisions, 12, 13, 47, 61
defense(s), 72, 73
definition, 62, 65
delivery, 38
demand, vii, viii, 1, 2, 7, 8, 10, 11, 14, 15, 16, 33, 35, 37, 41, 43, 47, 51, 54, 57, 58, 60, 63, 66, 76
denial, 16, 33
density, 41, 48, 58, 61

Index

Department of Transportation (USDOT), 10, 20, 28, 39, 44, 45, 46, 67, 68, 71, 72, 73, 74, 75, 76, 77, 78, 79
departments of transportation (DOTs), 12, 59, 60, 62
deregulation, 16, 17, 22, 23, 48, 49
direct measure, 45
distortions, 53
distribution, vii, 1, 2, 23, 30, 36
diversity, 35
downsizing, 63
duplication, 17

E

earth, 34
economic development, viii, 2, 3, 9, 29, 72
economic efficiency, 13
economic growth, vii, 1, 2, 3, 27
economic theory, 16
economy, 8, 9, 16, 53, 54
electronic, 26, 29
eligibility criteria, 29
employees, 22
employment, 15, 61
employment growth, 15
energy, 20, 26
energy efficiency, 26
engineers, viii, 2, 5, 34, 57
environment, 71
environmental, 3, 12, 13, 17, 20, 21, 53, 58, 59
environmental issues, 20
equilibrium, 16, 47
equity, 4
estimating, 37
evening, 39
evidence, 7, 16, 47, 48, 53
excess demand, 33, 34
expansions, 58
expert(s), 2, 7, 8, 11, 15, 23, 34, 36, 38, 49, 55, 58, 61, 65
external costs, 17

F

face-to-face interaction, 15
fairness, 9
family, 7
fear, 7, 14, 16
federal funds, 28
federal government, 3, 4, 11
Federal Highway Administration (FHWA), 9, 20, 35, 36, 39, 41, 42, 43, 44, 67, 68, 71, 72, 73, 74, 75, 76, 77, 79
federal law, vii, 1, 4, 9, 13, 48, 51
Federal Transit Administration, 46, 68, 76, 77
federal-aid, 3
Federal-Aid Highway Acts, 25
Federal-Aid Primary System, 26, 73
fee(s), 15, 17, 60, 61
finance, 17
financing, 3, 17, 27, 28, 29
firms, 17, 22
flexibility, viii, 2, 4, 9, 14, 26
flow, 7, 35, 36, 37, 59
focusing, 53
football, 38
fragmentation, 62
freight, vii, 1, 2, 4, 7, 8, 9, 11, 13, 14, 16, 22, 23, 27, 30, 33, 34, 35, 36, 41, 42, 43, 44, 47, 48, 49, 51, 52, 54, 57, 59, 63, 64, 66, 68, 69, 72, 73, 75, 77, 78
frustration, 65
fuel, 23, 53, 55, 78
full capacity, 23, 45
funding, vii, 2, 3, 4, 9, 10, 12, 14, 25, 26, 27, 29, 30, 31, 32, 62
funds, 3, 26, 27, 28, 29, 30, 32, 74

G

gasoline, 41, 76
gateways, vii, 1, 7, 12, 23, 43
General Accounting Office, 70

Georgia, 41, 62
Global Positioning System (GPS), 36
globalization, 41, 43
goals, 3, 12
government, viii, 2, 4, 9, 10, 11, 17, 47, 69
Government Accountability Office (GAO), 3, 67, 70, 71, 73, 77, 79
grants, 30
Great Depression, 19
gross domestic product (GDP), 9, 23, 40, 41, 54, 78
groups, 3
growth, 20, 22, 23, 25, 40, 41, 43, 57, 58, 61, 62, 72, 76
growth rate, 41

H

health, 16, 63
heavy rail, 21, 44, 45, 59
higher-income, 15
high-risk, 3
highway system, 9, 10, 11, 26, 27, 35, 36, 42, 72
Highway Trust Fund, 4, 9, 20
highways, 4, 10, 11, 14, 20, 27, 28, 29, 47, 48, 59, 63, 72, 74, 77
Homeland Security, 77
host, 10, 59, 62, 63
House, 26, 27, 30, 69, 73, 74, 79
household, 55
households, 41
housing, 61, 62
human, 72

I

implementation, 57
income(s), 8, 41, 61
Indiana, 11, 39, 72
Indiana East-West Toll Road, 11
indication, 48

indicators, 34, 35
indirect measure, 45
industrial, 41, 63
industry, 7, 11, 16, 17, 22, 23, 29, 47, 48, 49, 63, 72
inflation, 43, 54
information exchange, 29
information systems, 29, 59
infrastructure, 3, 10, 11, 12, 13, 14, 15, 17, 29, 30, 32, 33, 43, 47, 48, 63, 72
injuries, 3
innovation, 29
integration, 22
Intelligent Transportation Systems (ITS), 28, 29, 30, 32, 59, 79
intercity, vii, 1, 7, 8, 12, 41, 42, 44, 49, 51, 54
interference, 48
intermodal, 4, 28, 30, 33, 35, 47, 53, 54, 64
Intermodal Surface Transportation Efficiency Act (ISTEA), 9, 25, 26, 27, 28, 29, 31, 74
international, vii, 1, 2, 7, 43
international trade, vii, 1, 2, 7, 43
interstate, 21, 25, 26, 32, 57, 62, 72, 73
Interstate Highway Program, 19
Interstate system, 3, 9
inventories, 23
investment, 10, 11, 13, 17, 63

J

jobs, 8, 34, 38, 40, 58, 61, 62
jurisdiction, 62

K

Kentucky, 39

L

labor, 8, 54, 58, 59

Index

labor markets, 8
labor productivity, 54
land, 14, 16, 34, 43, 60, 61
land-use, 14, 34, 60, 61
law(s), 13, 17, 21, 22, 31
lead, 13, 15, 26, 37, 54, 62
legislation, 32
leisure, 21
level of service, 34
linear, 65
links, 27, 42
loans, 32
local government, 3, 11, 12, 13, 14, 19, 61, 62
location, 12
logistics, 22, 23
London, 60, 70, 75, 79
long distance, 48, 52
Los Angeles, 40, 45, 51
lower prices, 22
low-income, 61

M

macroeconomic, 11
maintenance, 10, 13, 15, 21, 26, 31, 60
major cities, 19, 20
major metropolitan areas, vii, 1, 4, 7, 12, 37, 58
management, 27, 28, 36, 59, 60, 76
Manhattan, 34
manufacturing, 36
marginal costs, 55
market(s), 47, 54, 63
Massachusetts, 69
measurement, 14
measures, vii, 2, 4, 14, 34, 35, 36, 48
media, 70
merchandise, 23, 43, 49
metric, 51, 72
Metropolitan planning organizations (MPOs), 12, 28
Mexico, 43
Miami, 45

Minnesota, 68
Mississippi River, 51
mobility, vii, 1, 3, 12, 15, 17, 26, 28, 29, 34, 35, 43, 59, 66, 67, 74
Model T, 19
models, 36, 37
money, 4, 11
morning, 37, 39
movement, 8, 54, 63, 79
Moynihan, Daniel, 25
music, 38

N

National Congestion Strategy, 2
National Highway System (NHS), 26, 28, 74
national income, 40
national security, 12
Nebraska, 8
network, 8, 15, 47, 48, 51, 59
New Orleans, 73
New York, 21, 45, 58
nodes, 4, 64
noise, 8, 64
North Carolina, 41
Northeast Corridor (NEC), 51, 52

O

Ohio, 13
oil, 20
Oregon, 36
organization(s), 2, 4, 8, 12, 28, 62, 64
ownership, 19, 21, 25, 71

P

Pacific, 7, 43, 49
paradox, 34
partnerships, 11
peak-period, 7, 8, 15, 16, 21, 33, 35, 37, 39, 40, 41, 42, 55, 58

pedestrian, 27
penalties, 36
performance, 10, 11, 14, 34, 51, 52
permit, 30
personal, 8, 21, 41
petroleum, 76
Philadelphia, 45
planners, viii, 2, 5, 57, 65, 66
planning, 4, 9, 12, 17, 26, 27, 34, 62
police, 62
policy makers, 5, 25, 57, 65
pollutants, 27
pollution, 26, 27, 64
poor, 15, 23, 38, 59
population, vii, 1, 2, 8, 9, 13, 15, 30, 39, 40, 41, 58
population growth, 41
population size, 13
ports, 4, 7, 23, 30, 41, 43, 64
power, 13, 62
pressure, 12, 15, 23, 43
price deflator, 78
prices, 11, 16, 22, 47, 51
priorities, vii, 2, 4, 9, 13, 32
private benefits, 64
private industry, 11
private investment, 11, 14
private sector, 11
privatization, 11
production, 19, 23, 43, 54
productivity, 8, 12, 16, 22, 23, 48, 53
productivity growth, 48
profitability, 23
profits, 11
program, 4, 9, 25, 26, 27, 28, 29, 30, 32, 36
programming, 4, 62
promote, 62, 72
proposition, 12
prosperity, vii, 1, 2, 3
public, vii, 2, 4, 11, 14, 16, 17, 21, 27, 32, 33, 34, 35, 36, 41, 44, 45, 54, 58, 62, 63, 64, 65
public funding, 63

public investment, 14
public policy, 17, 34, 65
public sector, 63
public-private, 11

R

rail, 4, 7, 11, 14, 16, 17, 21, 22, 23, 30, 32, 33, 35, 43, 44, 45, 46, 47, 48, 49, 51, 52, 53, 54, 57, 58, 59, 62, 63, 64
railroad, vii, 1, 2, 11, 16, 22, 23, 33, 38, 47, 48, 63, 72
range, 57, 60, 63
raw materials, 58
real terms, 9, 10, 40, 41, 49
reality, 37, 55
recognition, 25, 26
recreation, 34
recreational, 8
reduction, vii, 2, 4, 14, 42, 53, 57, 58, 61, 63
redundancy, 45
regional, vii, 1, 2, 12, 13, 15, 62
regional problem, 13
regulation(s), 17, 22, 72
relationship(s), 12, 36, 37, 47
relative size, 9
relevance, 28
reliability, 35, 36, 42, 43
repo, 68
resources, viii, 2, 3, 9, 10, 11, 12, 16, 17, 19, 21, 26
restaurant(s), 8, 34
retail, 34
returns, 13
revenue, 16, 48, 49, 61, 72
rights-of-way, 58
rings, 41
risk, 48
rivers, 15, 61
road(s), vii, 1, 2, 4, 8, 13, 14, 15, 19, 20, 21, 23, 25, 26, 27, 33, 34, 35, 36, 37, 38, 39, 40, 41, 42, 43, 44, 51, 53, 54, 57, 58, 59, 60, 61, 63, 71

routing, 22, 48
rural areas, 11, 13, 25, 27, 42, 45, 76

S

Safe, Accountable, Flexible, Efficient Transportation Equity Act (SAFETEA), vii, 1, 2, 3, 10, 30, 31, 32, 74
safety, viii, 2, 9, 12, 17, 20, 26, 29, 30, 53, 79
scarce resources, 26, 66
school, 36, 38
seaports, vii, 1, 2
Seattle, 45
Second World War, 19
Secretary of Transportation, 2
security, 3, 43
Senate, 26, 73, 77
separation, 17, 44
series, 43
service quality, 15, 33, 47
severity, 53
shareholders, 11
shocks, 20
sign, 8
signals, 59, 63
social environment, 64
social situations, 15
society, 37
Southern Pacific, 7, 49
speech, 72
speed, 30, 33, 34, 35, 36, 37, 46, 48, 49, 51, 63, 64
speed limit, 37
stability, 3
staffing, 43
stakeholder(s), 9, 13, 27
standards, 29
statistics, 77
stop-and-go, 33
Strategic Highway Network, 26, 73
strategies, 13, 16, 57, 59, 60, 61, 62, 66
strength, 16

stress, 33, 53, 54
suburbs, 40, 58
summer, 43
supply, 8, 11, 15, 16, 23, 34, 35, 47, 60, 66
supply chain, 8, 23, 34
surface transportation, viii, 4, 7, 25, 26
Surface Transportation Board, 72
Surface Transportation Program (STP), 26, 27, 31
switching, 27
synthesis, 75, 77
systems, vii, 1, 2, 4, 7, 10, 19, 21, 22, 27, 44, 45, 46, 47, 51, 58, 59, 62, 63, 64, 66

T

taxation, 16, 61
technological change, 22, 63
technology, 28, 36, 48, 60
tension, 3
terminals, 8, 23, 47
terrorist attack, 43
Texas, 7, 35, 39, 40, 41, 67, 70, 74, 75, 76, 77, 79
Texas Transportation Institute (TTI), 7, 8, 9, 35, 36, 37, 38, 39, 40, 42, 53, 55, 57, 58, 67, 70, 74, 75, 76, 77, 78, 79
theoretical, 35, 36
theory, 14, 48
thinking, 55
threat(s), 2, 19
time, 3, 7, 16, 17, 19, 20, 21, 22, 26, 32, 34, 35, 36, 37, 39, 41, 43, 45, 46, 47, 51, 52, 53, 55, 58, 59, 60, 61, 63, 65, 73, 78
timing, 38, 48
Title III, 31
tolls, 26, 30, 32, 60, 61
total costs, 55
tracking, 60
trade, vii, 1, 7, 12, 23, 43

traffic, vii, 2, 4, 8, 13, 14, 15, 19, 20, 21, 23, 25, 26, 27, 32, 33, 34, 35, 36, 37, 38, 39, 40, 41, 42, 43, 44, 48, 49, 51, 53, 54, 55, 58, 59, 60, 61, 63, 72
traffic flow, 27
transport, 47, 68
transportation, vii, viii, 1, 2, 3, 4, 7, 8, 9, 10, 11, 12, 13, 14, 15, 16, 19, 20, 21, 22, 23, 25, 26, 27, 30, 31, 33, 34, 35, 41, 43, 45, 47, 48, 53, 54, 57, 58, 59, 60, 61, 62, 64, 65, 69, 77
Transportation Research Board (TRB), 2, 48, 67, 70, 71, 73, 74, 76, 77, 78
trend, 49
trucking, 17, 22, 23, 54, 63
trucks, vii, 1, 3, 4, 8, 16, 36, 42, 43, 63
turbulent, 49

U

unit cost, 54
United Kingdom, 13
United States, 3, 4, 8, 19, 40, 43, 44, 45, 68, 73, 75, 76
UPS, 34
urban areas, 7, 9, 11, 13, 15, 16, 23, 27, 38, 39, 40, 41, 42, 45, 47, 54, 58, 62, 64, 76, 78
urban centers, 23
urban sprawl, vii, 1, 8
urbanized, 30

users, vii, 1, 2, 4, 16, 58, 61

V

value pricing, 15
variability, 35, 43
variable(s), 30, 36, 46, 47, 53, 60
vehicles, 15, 19, 21, 35, 37, 38, 40, 41, 45, 46, 53, 58, 61
vein, 2
Virginia, 41
voting, 13

W

wait times, 43
walking, 61
war, 19
warrants, 4
Washington, 20, 21, 24, 36, 38, 44, 45, 46, 49, 50, 67, 68, 69, 70, 71, 72, 73, 74, 75, 76, 77, 78, 79
water, 22, 43, 64
wear, 63
welfare, 3
wells, 79
workers, 8, 42, 54, 58
worry, 51